ANNALS *of* THE NEW YORK ACADEMY OF SCIENCES

VOLUME
1292

ISBN-10: 1-57331-931-7; **ISBN-13:** 978-1-57331-931-7

ISSUE

Annals Reports

The conference "Play, Attention, and Learning: How Do Play and Timing Shape the Development of Attention and Facilitate Classroom Learning?," held on June 15, 2012, was presented by the New York Academy of Sciences and the Brain Trauma Foundation, and supported by educational grants from the Leon Levy Foundation and NewYork-Presbyterian Hospital.

The meeting "Treatment-Resistant Depression: Glutamate, Stress Hormones, and their Roles in the Regeneration of Neurons" was held on March 25, 2013 at the New York Academy of Sciences and presented by the Academy's Biochemical Pharmacology Discussion Group, which is sponsored by the American Chemical Society, Boehringer Ingelheim, and Pfizer.

TABLE OF CONTENTS

T0188535

Annals of the New York Academy of Sciences (ISSN: 0077-8923 [print]; ISSN: 1749-6632 [online]) is published 30 times a year on behalf of the New York Academy of Sciences by Wiley Subscription Services, Inc., a Wiley Company, 111 River Street, Hoboken, NJ 07030-5774.

Mailing: *Annals of the New York Academy of Sciences* is mailed standard rate.

Postmaster: Send all address changes to ANNALS OF THE NEW YORK ACADEMY OF SCIENCES, Journal Customer Services, John Wiley & Sons Inc., 350 Main Street, Malden, MA 02148-5020.

Disclaimer: The publisher, the New York Academy of Sciences, and the editors cannot be held responsible for errors or any consequences arising from the use of information contained in this publication; the views and opinions expressed do not necessarily reflect those of the publisher, the New York Academy of Sciences, and editors, neither does the publication of advertisements constitute any endorsement by the publisher, the New York Academy of Sciences and editors of the products advertised.

Publisher: *Annals of the New York Academy of Sciences* is published by Wiley Periodicals, Inc., Commerce Place, 350 Main Street, Malden, MA 02148; Telephone: 781 388 8200; Fax: 781 388 8210.

Journal Customer Services: For ordering information, claims, and any inquiry concerning your subscription, please go to www.wileycustomerhelp.com/ask or contact your nearest office. *Americas:* Email: cs-journals@wiley.com; Tel:+1 781 388 8598 or 1 800 835 6770 (Toll free in the USA & Canada). *Europe, Middle East, Asia:* Email: cs-journals@wiley. com; Tel: +44 (0) 1865 778315. *Asia Pacific:* Email: cs-journals@wiley.com; Tel: +65 6511 8000. *Japan:* For Japanese speaking support, Email: cs-japan@wiley.com; Tel: +65 6511 8010 or Tel (toll-free): 005 316 50 480. Visit our Online Customer Get-Help available in 6 languages at www.wileycustomerhelp.com.

Information for Subscribers: *Annals of the New York Academy of Sciences* is published in 30 volumes per year. Subscription prices for 2013 are: Print & Online: US$6,053 (US), US$6,589 (Rest of World), €4,269 (Europe), £3,364 (UK). Prices are exclusive of tax. Australian GST, Canadian GST, and European VAT will be applied at the appropriate rates. For more information on current tax rates, please go to www.wileyonlinelibrary.com/tax-vat. The price includes online access to the current and all online back files to January 1, 2009, where available. For other pricing options, including access information and terms and conditions, please visit www.wileyonlinelibrary.com/access.

Delivery Terms and Legal Title: Where the subscription price includes print volumes and delivery is to the recipient's address, delivery terms are Delivered at Place (DAP); the recipient is responsible for paying any import duty or taxes. Title to all volumes transfers FOB our shipping point, freight prepaid. We will endeavour to fulfill claims for missing or damaged copies within six months of publication, within our reasonable discretion and subject to availability.

Back issues: Recent single volumes are available to institutions at the current single volume price from cs-journals@wiley.com. Earlier volumes may be obtained from Periodicals Service Company, 11 Main Street, Germantown, NY 12526, USA. Tel: +1 518 537 4700, Fax: +1 518 537 5899, Email: psc@periodicals.com. For submission instructions, subscription, and all other information visit: www.wileyonlinelibrary.com/journal/nyas.

Production Editors: Kelly McSweeney and Allie Struzik (email: nyas@wiley.com).

Commercial Reprints: Dan Nicholas (email: dnicholas@wiley.com).

Membership information: Members may order copies of *Annals* volumes directly from the Academy by visiting www. nyas.org/annals, emailing customerservice@nyas.org, faxing +1 212 298 3650, or calling 1 800 843 6927 (toll free in the USA), or +1 212 298 8640. For more information on becoming a member of the New York Academy of Sciences, please visit www.nyas.org/membership. Claims and inquiries on member orders should be directed to the Academy at email: membership@nyas.org or Tel: 1 800 843 6927 (toll free in the USA) or +1 212 298 8640.

Printed in the USA by The Sheridan Group.

View *Annals* online at www.wileyonlinelibrary.com/journal/nyas.

Abstracting and Indexing Services: *Annals of the New York Academy of Sciences* is indexed by MEDLINE, Science Citation Index, and SCOPUS. For a complete list of A&I services, please visit the journal homepage at www. wileyonlinelibrary.com/journal/nyas.

Access to *Annals* is available free online within institutions in the developing world through the AGORA initiative with the FAO, the HINARI initiative with the WHO, and the OARE initiative with UNEP. For information, visit www. aginternetwork.org, www.healthinternetwork.org, www.oarescience.org.

Annals of the New York Academy of Sciences accepts articles for Open Access publication. Please visit http://olabout.wiley.com/WileyCDA/Section/id-406241.html for further information about OnlineOpen.

Wiley's Corporate Citizenship initiative seeks to address the environmental, social, economic, and ethical challenges faced in our business and which are important to our diverse stakeholder groups. Since launching the initiative, we have focused on sharing our content with those in need, enhancing community philanthropy, reducing our carbon impact, creating global guidelines and best practices for paper use, establishing a vendor code of ethics, and engaging our colleagues and other stakeholders in our efforts. Follow our progress at www.wiley.com/go/citizenship.

Ann. N.Y. Acad. Sci. ISSN 0077-8923

ANNALS OF THE NEW YORK ACADEMY OF SCIENCES

Issue: Annals *Reports*

Play, attention, and learning: How do play and timing shape the development of attention and influence classroom learning?

James H. Hedges,[1] Karen E. Adolph,[2] Dima Amso,[3] Daphne Bavelier,[4,5,6] Julie A. Fiez,[7] Leah Krubitzer,[8,9] J. Devin McAuley,[10] Nora S. Newcombe,[11] Susan M. Fitzpatrick,[12] and Jamshid Ghajar[1,13]

[1]Brain Trauma Foundation, New York, New York. [2]Department of Psychology, New York University, New York, New York. [3]Department of Cognitive, Linguistic, and Psychological Sciences, Brown University, Providence, Rhode Island. [4]Rochester Center for Brain Imaging, Rochester, New York. [5]Department of Brain and Cognitive Sciences, University of Rochester, Rochester, New York. [6]Faculty of Psychology and Educational Sciences, University of Geneva, Geneva, Switzerland. [7]Center for Neuroscience, University of Pittsburgh, Pittsburgh, Pennsylvania. [8]Department of Psychology, University of California, Davis, Davis, California. [9]Center for Neuroscience, University of California, Davis, Davis, California. [10]Department of Psychology, Michigan State University, East Lansing, Michigan. [11]Department of Psychology, Temple University, Philadelphia, Pennsylvania. [12]James S. McDonnell Foundation, St. Louis, Missouri. [13]Department of Neurological Surgery, Weill Cornell Medical College, New York, New York

Address for correspondence: Jamshid Ghajar, Brain Trauma Foundation, 7 World Trade Center, 34th Floor, 250 Greenwich St., New York, NY 10007. jam@ghajar.net

The behavioral and neurobiological connections between play and the development of critical cognitive functions, such as attention, remain largely unknown. We do not yet know how these connections relate to the formation of specific abilities, such as spatial ability, and to learning in formal environments, such as in the classroom. Insights into these issues would be beneficial not only for understanding play, attention, and learning individually, but also for the development of more efficacious systems for learning and for the treatment of neurodevelopmental disorders. Different operational definitions of play can incorporate or exclude varying types of behavior, emphasize varying developmental time points, and motivate different research questions. Relevant questions to be explored in this area include, How do particular kinds of play relate to the development of particular kinds of abilities later in life? How does play vary across societies and species in the context of evolution? Does play facilitate a shift from reactive to predictive timing, and is its connection to timing unique or particularly significant? This report will outline important research steps that need to be taken in order to address these and other questions about play, human activity, and cognitive functions.

Keywords: play; attention; learning; education; anticipatory timing; synchronization; infant development; locomotion; perceptual–motor coordination; action video games; architecture; isotropic fractionator; head-mounted eye-tracking; cortex; evolution; spatial skill; puzzles; child ANT; transfer; STEM

Overview of the workshop

"Play, Attention, and Learning: How Do Play and Timing Shape the Development of Attention and Facilitate Classroom Learning?" was a one-day workshop convened by the New York Academy of Sciences and the Brain Trauma Foundation on June 15, 2012 in New York City. The workshop explored the idea that the design of classroom-based learning activities implicitly builds on many of the cognitive abilities children typically acquire through informal activities earlier in childhood, including physical play. Neural connections that facilitate synchronizing temporal and spatial expectancy with incoming sensory information may be formed through certain activities requiring children to clap, hop, or perform other rhythmic actions dependent on anticipation and timing.

doi: 10.1111/nyas.12154

Ann. N.Y. Acad. Sci. 1292 (2013) 1–20 © 2013 The Authors. *Annals of the New York Academy of Sciences*
published by Wiley Periodicals Inc. on behalf of The New York Academy of Sciences.

1

Hypothesizing a connection between play and the development of important cognitive abilities expands the notion that interacting with the environment—particularly when engaged in activities that rely on anticipatory timing, cadence, or actions linked to subsequent actions—shapes the development of the attention network. Engaging in play could provide the developing brain with spatially and temporally predictive interactions with the outside world and thereby tune the developing network's ability to select which information to attend to when, and which information to ignore.

The main goal of the workshop was to review the current state of scientific knowledge and to make recommendations for future research priorities. The role of timing in play and the development of attention has not been a traditional focus in developmental neuroscience research. With recent interest in the role of predictive timing in attentional focus, the natural progression is to ask how this capacity develops and whether it has an impact on subsequent learning ability.

The workshop was organized into two sessions. In the first session, a series of overview lectures provided the participants with an introduction to the current state of knowledge about the role of play in children's cognitive development from the perspective of different disciplines and experimental approaches. The second session consisted of breakout discussion groups charged with reviewing existing knowledge on issues posed by the workshop organizers (detailed below), brainstorming and identifying areas of converging research, developing possibilities for future work, and considering how these activities may aid in the design of interventions for children with attention-related and learning disabilities.

Introductory lectures

Jamshid Ghajar (The Brain Trauma Foundation) gave the first of two brief introductory talks preceding the overview lectures. He described the genesis of the workshop and discussed research conducted by the Brain Trauma Foundation in people diagnosed with concussion, with an interest in the connection between attention and predictive timing. In collaboration with Richard Ivry at the University of California, Berkeley, he tested the hypothesis that predictive timing is an essential element of attention in dynamic interactions.[1] This led to the ques-

Figure 1. Tennis player, Roger Federer, is shown during a forehand stroke. Comparison of the latencies for sensory awareness and movement execution to the speed of the incoming ball highlights the necessity of predictive timing. The absence of accurate predictions for these varying delays would obviate making contact with the ball.

tion of how predictive timing, and thereby attention, develops.

Synchronization with the outside world in dynamic interactions requires accurate predictive timing of the to-be-attended-to sensory information. It is important to be able to predict when relevant sensory events that are necessary for appropriate behavior will occur. Consider a tennis player who wants to hit a fast-moving ball but has varying latencies in sensory and motor processing (Fig. 1). To make contact with the ball, the player needs to predict when and where the ball is going to land, based, in part, on the motor actions of his opponent hitting the ball. Without the ability to predict when and where the ball will land, he would need to repeatedly swing his racquet to guess when the ball will arrive and would need to expend considerable resources to do so.

The predictive timing required to perform this racquet swing is analogous to that required in cognitive processing: to selectively attend, one needs to predict when the sensory information arrives, and which inputs are relevant to the situation, so that

processing can occur efficiently and appropriate behavior can be generated. To make the connection
between predictive timing and learning, consider a
teacher and students in a classroom. To process the
teacher's words, the students need to predict the timing or the cadence of the teacher's speech, allowing
them to process the speech content just in time. Any
deficiency in this kind of prediction could significantly impair their ability to listen and remember,
which can manifest as a learning disability.

Following these points, Ghajar considered what
it means to play (i.e., how it can be defined). In his
view, play may be a biological activity within which
predictive timing develops. Play during early childhood coincides with cerebellar granule cell migration and synaptogenesis, and since the cerebellum
has a known role in predictive timing, play may be
the key to the development of this ability. Young
children seem to seek out predictable interactions
and then endlessly repeat them. By example, consider a young boy who repeatedly throws stones into
a puddle (Fig. 2). He releases the stone, and after
a certain period of time, there is a splash. He re-
peats this action until the expectancy of the splash
matches the actual timing of the splash. This may
be the result not only of reducing the variability in
his motor process for throwing the stone, but also
of forming better predictions of the stone's spatial
and temporal dynamics.

Following Ghajar, Susan Fitzpatrick (the James
S. McDonnell Foundation) began her presentation
with examples of the universality of certain aspects
of play. Some childhood behaviors, such as playing patty-cake, have ancient roots and are common
in many cultures around the world. These observations have led Fitzpatrick and others to speculate
on whether some of the experiences common to
play contribute to the abilities children call on for
subsequent learning. It may be that some kinds of
childhood activities should be considered a form of
species-typical behavior necessary for the development of some of the brain networks required for suc-
cessfully building the skills necessary for classroom
learning. A first step in exploring these possibilities
in future studies might be to determine the extent
to which children engage in similar behaviors across
cultures and whether play contributes to cognitive
development.

Fitzpatrick is particularly intrigued with the many
types of childhood play that are dependent upon

Figure 2. A boy is shown dropping a stone into a puddle.
By repeating this predictable activity, he may develop stored
representations of the properties of the external world from
which accurate predictions of those properties can be formed.

timing. Consider the hand-clapping game "Mary
Mack," where there is an element of prediction:
movements must be synchronized so that hands
are in the right place at the right time. The game
also builds on rhyming and song; children stand or
sit opposite one another and clap hands in tune to
the song. Timing is similarly important in "Double
Dutch," a game in which one or more players simultaneously jump over two long jump ropes that
are turned in opposite directions. Predictive timing
is key to jumping rope. In order to avoid getting hit
by the rope, the jumper needs to know when it is
going to arrive and how long it will take to execute
his/her jump.

Fitzpatrick closed her talk by posing additional
questions on how play could influence outcomes
later in life. What might be the expected outcomes
for children who do not engage in species-typical
behaviors? When children are unable, for whatever

reasons, to take part in the informal activities that might contribute to shaping the brain networks called upon in more formal learning situations, such as in a classroom, what kinds of activities could constitute this crucial form of play? What are the necessary elements? Could some children, struggling to make the transitions in early education, benefit from play-based interventions? Is it possible that the way skills are developed informally could be co-opted as a tool for building the skills required for success in more formal learning environments?

Overview lectures

Development of movement

Karen E. Adolph (New York University) initiated the expert overviews with a presentation entitled "Play and Human Development." Adolph's first point was that play occupies an immense part of an infant's daily activity. Play—with objects, people, and features of the environment—generates a wealth of information. What are the opportunities for learning within play? How does the information that infants generate during play lead to learning? What kinds of information do infants take in and what are the properties of that information? Adolph and colleagues have worked to answer these questions within the context of locomotor, object, and social play.

The play of a typical toddler involves many different activities simultaneously: looking around, walking, holding objects, and interacting socially with others in the room. As Adolph showed in a series of video exemplars, head-mounted eye tracking combined with video tracking reveals a complex set of overlapping activities, with quick switches within and between activities, changes in speed, and continual starts and stops in locomotion.[2,3] Eye gaze frequently switches among the targets of these simultaneous forms of varied activity (e.g., from a ball on the floor, to an upcoming obstacle, to a caregiver's hands and feet); infants may pause to focus on the yellow ball or the doll in the scene.

Locomotor play is also extremely varied. Typically, infants' movements cover an entire room, all the while engaged in different activities and looking at many different things.[4] A graphic rendition of 10 minutes of spontaneous toddler activity reveals a twisting path that covers most of the locations in the room, with repeated loops through locations of interest. A raster plot of spontaneous walking from 60 infants confirms what we see in the video exemplars: locomotion is distributed in short bursts of activity, with longer periods of rest in between; each burst occurs in a different physical and social context.[4] Infants' spontaneous play can be described as "repetition without repetition," an expression of the late Soviet neurophysiologist, Nikolai Bernstein.

Taken together, the above observations emphasize that any theory of learning has to work on multiple simultaneous streams of input. The relationships among the morphological characteristics of infants' bodies, motor skills, incoming sensory information, and subsequent learning are multidirectional and dynamic. Methodological biases, such as an overreliance on looking measures, may obscure the complexity of these issues and may divert attention away from the extraordinary richness of babies' behaviors.

Adolph argued that researchers should not rely on visual observation alone or assume that they know what features of the behavior are important. Rather, they should objectively collect and measure data at a number of levels, from eye tracking to walking. A case in point: using head-mounted eye tracking during free play to determine whether babies actually look, and how often, at their mothers. Although most researchers assume that infants look to their mothers' faces, the data show that, in fact, infants look at their mothers only 54% of the time following mothers' vocalizations; nearly half the time, they continue looking at whatever they are already looking at.[2] When infants do look at their mothers, they rarely look at their faces and instead look at their bodies or hands. These findings were far more likely to be revealed through the new technology of head-mounted eye tracking than with visual observation alone.

Spontaneous locomotor activity increases sharply from experienced crawlers to novice walkers, and continues to increase in walkers from 12 and 19 months of age. Toddlers walk a lot; the average toddler takes 2368 steps/hour, which corresponds to roughly 14,000 steps/day.[4] They can travel immense distances, about 701 m/hour, or the equivalent of 46 football fields/day. Object play is similarly varied, distributed, and immense. They touch many different objects, often all at once. Infants are in contact with objects for more than 30 minutes out of every hour, and there is a significant increase in object play between the ages of 11 and 13 months.[5]

Ann. N.Y. Acad. Sci. 1292 (2013) 1–20 © 2013 The Authors. *Annals of the New York Academy of Sciences*
published by Wiley Periodicals Inc. on behalf of The New York Academy of Sciences.

The transition from crawling to walking affects what infants see. Thus, changes in motor skill lead to changes in play and corresponding changes in the input. While crawling, infants keep their heads down and look at the floor; people, toys, objects, and the facing walls are out of view. While walking, the whole room comes into view.[6] The transition from crawling to walking also has a significant impact on object carrying.[7] Walking infants carry objects much more than crawlers do (e.g., one walker carried objects 144 times within one hour). And although crawling infants also can carry objects (e.g., one crawler carried objects 20 times within one hour), they have a different way of doing it. Crawlers crawl while holding an object (e.g., under their arm or in their mouth) or while pushing it, or they bum-shuffle while holding an object in their hands.

Developmental changes in posture affect how infants share objects with their mothers and, more generally, how infants and mothers interact.[5] Social interactions emerge from object carrying. Do infants carry objects to share them with their mothers? Yes they do. Crawlers share objects with their mothers, but primarily from a stationary position. They typically sit in one place holding up an object, and the mother has to come over to them to engage in the interaction. In contrast, walkers share objects while mobile. They pick up the object and carry it to the mother to engage in the interaction.

Differences in how objects are shared matter to mothers. Developmental changes in posture initiate a developmental cascade, such that mothers' verbal responses to infants depend on how infants share.[8] When infants sit in one place and try to share an object by holding it up, mothers typically respond by ignoring infants' bids or affirming what they did, such as by saying "thank you;" less frequently, they name the object, such as by saying "oh, an orange ball;" sometimes they give action directives, such as by saying "put the block in first." In contrast, when walkers carry objects to their mothers, the mothers respond with more action directives than any other type of response. It should be noted though that the difference in maternal responding is not merely due to an infant's upright posture. In the few cases when crawlers carried objects to their mothers, the mothers responded primarily with action directives. In other words, postural development affects how infants share objects, and how they share objects, in turn, affects how mothers respond.

A similar cascade of developmental changes relates to 3-D form perception—knowing what the backside of an object is before you turn it around. Since objects are self-occluding, one needs to know what should be on the other side to perceive its 3-D form. A straightforward way to test 3-D form perception in infants is to habituate them to the frontal view of an object by rotating it a few degrees from side to side, and then test whether they look longer at an incomplete rather than complete version of the object when it is rotated 360°. If infants look longer at the incomplete version, it can be inferred that they perceived the full 3-D form during habituation.[9]

Applying this task, one finds that the ability to sit independently predicts 3-D form perception. Why might this be the case? Sitting frees infants' hands, allowing them to rotate objects, transfer objects from hand to hand, and finger the edges of objects. These spontaneous play activities support the acquisition of a fuller representation of object form. In short, independent sitting gives infants more opportunities for visual–manual object exploration. Infants who cannot yet sit rotate objects in the same frequency with and without looking at them, whereas infants who can sit do much more exploration with looking. The ability to explore objects by looking and touching leads to more sophisticated object knowledge.[9]

Adolph closed by reviewing the central points of her talk: play is repetition without repetition; infant play is complex and involves switching between multiple parallel streams; developmental changes (e.g., in posture, body growth, or skill) shape how infants play, which thereby facilitates access to information. Before taking questions, Adolph answered one of her own: Why do babies play?—because they can, because it is fun, and because it is interesting.

Fitzpatrick asked what happens if infants cannot play. Adolph said that there is much redundancy to increase the likelihood that they will be able to, but that if they cannot, they will not generate as much information for themselves, and opportunities for learning will be curtailed. Leah Krubitzer (University of California, Davis) commented that play is the only thing that infants do and wondered whether adults continue to play as well. Adolph responded by pointing out that adults do continue to play, but that much of infants' play is novel. Devin McAuley (Michigan State University) asked about 3-D form perception and whether it is important that babies are actually rotating the objects. Adolph answered that passive exposure to objects does not lead to the development of 3-D–object form perception; rotation is indeed critical.

Development of attention

Bruce McCandliss (Vanderbilt University) gave the second overview lecture. He began by noting a delightful quote from Albert Einstein: "Understanding physics is child's play when compared to understanding child's play." McCandliss pointed out that you could replace the second mention of child's play in this quote with attention, since, in his view, attention is an incredibly complicated phenomenon. McCandliss presented a series of questions: How do attentional networks relate to play? How do we learn to attend? How does that change over the course of development?—which prompts the question of what matters in life experience. How would play impact these issues? Assuming play is important to the development of attention, how can we get more play into education?

To answer these questions, one could isolate an attention-related activity in someone's behavioral repertoire (e.g., how well they do in attending to something) and connect that to changes in their brain networks—exactly the approach that some labs have taken. McCandliss pointed out that Michael Posner and Steven Petersen, then at Washington University in St. Louis, wrote an influential paper about attention that looked at how brain damage relates to particular deficits in attention.[10] They proposed the idea that there might be different networks relating to different subsystems of attention (i.e., different syndromes relating to different patterns of damage). They identified three, somewhat distinct, subsystems of attention: alerting, defined as achieving a state of readiness; orienting, defined as the selection of information from sensory input; and executive attention, defined as detecting and resolving conflict between potential responses.[11]

McCandliss continued by noting that a further innovation by Posner and colleagues came by developing a paradigm, the Attention Network Task (ANT), that could be used to look at the principles of attention by pushing around a simple decision.[12] The decision in the ANT is based on whether something is pointing one way or another, akin to judging the directions of pointed arrows in the arrow-based

version of the Eriksen flanker task.[13] By applying different cue and target conditions, the ANT allows for the quantification of orienting, alerting, and executive attention. Measures of each are based on how long it takes the brain to process different information (i.e., by comparing reaction times).

An alerting cue can decrease the amount of time it takes the brain to process information, in general, by about 50 milliseconds.[12] An orienting cue can also accelerate information processing by facilitating the allocation of attention to a particular region of visual space, which lowers the time it takes to process a target in that region.[12] A set of targets that conflict, by contrast, can slow things down, perhaps by demanding additional processing to resolve the conflict.[12] One way to think of the latter is in terms of driving. If a global positioning system tells the driver to go one way and a passenger in the car recommends going another way, it can take additional time to process that kind of discordant information.

McCandliss pointed out that the ANT has been adapted for young children by replacing the standard arrows used for adults with oriented cartoon-like fish.[14] Results from the child version of the ANT revealed a developmental progression for alerting and orienting attention between four and seven years of age.[14] This progression parallels the development of the frontal system, which enables greater cognitive control. Conflicting information, however, still generates significant problems for children in this age group.[14] It can result in a doubling in reaction times and a fivefold increase in errors. Related imaging work suggests that conflict is represented in the anterior cingulate cortex (ACC) and the dorsolateral prefrontal cortex (DLPFC),[15] structures that are late to develop and, in cases of parental conflict, can cause greater challenges in adults. Recent work has also explored how event-related potentials (ERPs) relate to developmental changes in young children.[14] These results suggest that children's brains are more sensitive to conflict, and that this sensitivity is progressively diminished as they develop into adulthood.

McCandliss referred to work by Adele Diamond showing that children can deal better with conflict if they engage in structured activities, especially those augmented with contextual information on the fly.[16,17] Being engaged in something often leads to anticipation of when it is going to happen. The neural correlates of such anticipation consist of overlapping brain networks that are activated during the cue period of attention tasks (i.e., between the presentation of the cue and the presentation of the target).[18] Different regions are activated for *getting ready* and for *handling conflict*. Anticipating an action that might have a conflict associated with it evokes activity in the presupplementary motor area, which is thought to be involved in creating a copy of what you are going to see and what is going to happen. Such copies are continuously compared in a flow state where people can adapt their behavior. Regions of the basal ganglia that facilitate millisecond timing are also thought to be involved.

In addition, contingent negative variation, an ERP component, has been shown to relate to expectancy.[19] It can be used to predict how quickly someone is going to react to something, which therefore suggests that it is capturing some aspect of attention. McCandliss suggested that measuring contingent negative variation could be one way of identifying the signature of a shift from more reactive to more predictive processing. He expressed the view that these aspects of the development of the temporal dynamics of attention are likely important and yet remain largely unexplored. There could be a shift in the ability to predict with play, where children become more predictive, but additional work is needed to test this possibility.

McCandliss pointed out that advanced go/no-go tasks, such as the AX version of the continuous performance tasks (AXCPTs), have been used as another window into children's cognition.[20] To correctly perform an AXCPT, a subject presses a key after a particular letter, such as X, which is presented after another particular letter, such as A. The subject presses a different key after the presentation of all other letters. The time between presentations of letters varies but is predictable. Performance on the AXCPT shows a developmental progression between five and nine years of age. Children who are proactively engaging in the world differ in their performance from those who are not. Nine-year-old children also show an expectancy wave—a prediction of when each letter is going to show up, whereas younger children do not. Other signs of predictive ability on this task, such as pupil dilation, show up around eight years of age.

McCandliss changed directions and noted that play can also have an element of cruelty. Cruelty has been seen in play in the virtual game Cyberball, which is based on an Atari arcade game of seven-man American football.[21] In some experimental settings, children play the game simultaneously while they are in a magnetic resonance imaging (MRI) scanner. They chose different teammates, a process that can make some of them feel excluded. During the exclusion phase, there is activation of the ACC that is more similar to activity seen during emotional conflict than during typical attentional activity. Activity in the subgenual ACC has been shown to be related to depression in 13-year-old girls.[22]

McCandliss ended by emphasizing that despite the above-mentioned insights, there is a period of development wherein little is known about attentional function, roughly between two and five years of age.

Alison Gopnik (University of California, Berkeley) cautioned that an assumption in the executive attention literature is that if successful adults do something, it must be good for children to do the same thing. Gopnik added that for the child version of the flanker task, if the focus is on getting the task right, then an observer would reach the conclusion that children do not perform well. But they may be trying to do something else, such as inferring a generalization of how many fish go in the same direction, and that they may be performing quite well, but doing so at the expense of the more adult-like objective of the task. In other words, infants may have the capacity to do something much more broad in relation to exploring their environment (than we might first think). Issues of this kind can be thought of as a trade-off between control and other factors, such as generalization. With this in mind, new tasks could be developed to push decision dynamics in new directions. Games could be developed that lack instructions, such as discovering as many fish as possible.

Learning to attend

Daphne Bavelier (University of Rochester and University of Geneva) gave the third overview lecture, entitled "Learning to Attend: Lessons from Action Videogames." Bavelier's interest is in learning and brain plasticity, which she explores in the context of action video games, and in particular, first- or third-person shooter games. Action video games may be thought of as quite mindless, consisting simply of avatars running around shooting the bad guy; but practicing them, in fact, imparts benefits on a wide range of cognitive tasks.[23] Playing action video games can have beneficial effects that include transfer effects to basic aspects of attention and to many other skills. They can, for example, affect visual attention and the efficiency of switching between tasks.

Bavelier listed a number of important questions: How does playing action video games affect how well one sees, or how well one can perform cognitive processes like mental rotation? And why does such a wide array of skills appear to be modified by playing these games? Are transfer effects a consequence of a general improvement in attentional control or, instead, the result of independent changes in each and every function? Toward answering these questions, Bavelier's working hypothesis is that playing action games improves top–down attentional control and, in doing so, hones the ability of players to differentiate signal from noise. As a result, gamers are able to carry out more accurate inferences in the service of decision making and thus show more skilled performance on a variety of different cognitive tasks.

Bavelier pointed out that her work follows the Bayesian brain approach in which the human nervous system has as its main task to infer the most likely way to react given the information perceived, the stated task goal, and previous knowledge.[24] Loosely speaking, this view holds that the brain at all times predicts what the best decision should be given the present state of the world and past experiences. For example, while someone is speaking, a listener's brain will be engaged in predicting what the speaker is going to say based on the speech stream heard up to a given point, as well as the context of the discussion. Individuals who can make more accurate predictions will zoom in on the most appropriate decision faster. Bavelier's lab has shown that, indeed, action video game play trains players to make more accurate decisions.

To illustrate how research in her lab proceeds, Bavelier presented a concise example from her work on changing vision for the better. Contrast sensitivity is the ability to distinguish small shades of gray; it is a fundamental aspect of vision that can, for example, make the difference between crashing and not crashing into the car in front while driving in a thick fog. Measuring contrast sensitivity with small sine-wave gratings, termed *Gabor patches*, is a common approach in vision science. The standard task consists of two brief presentations of small Gabor patches; the subject indicates whether the patch was there in the first or second presentation.

Bavelier conducted a study in which contrast sensitivity was measured before and after undergraduate students played video games for 50 h over the course of 10 weeks.[25] Critically, participants were randomly assigned to an action video or a control game; in both cases, the games were selected from commercially available options that are known to be quite enticing. A major difference in the two types of games is that the control game did not have the same level of dynamics, including fast pace, the need for visuomotor control and divided attention, as well as the meshing of many goals and subgoals at many different time scales. Bavelier confirmed that subjects who played action video games had higher contrast sensitivity than control trainees. She also alluded to related work by Ian Spence and Jay Pratt at the University of Toronto that employed a similar group design and that established that playing action games enhances mental rotation skills more than playing control games.

Transfer effects from playing action video games may be partly explained by changes in attentional control. Data from Bavelier and others show that playing them sharpens top–down attentional control, such as selective attention over space, time, and objects.[26,27] Bavelier and colleagues have found that habitual action video game players are better at suppressing distracting information and that the better the suppression, the faster the reaction time. This effect has been noted not just in young adults but also in children who engage in action video game play, although the details associated with the effects of playing at varying ages have yet to be explored. Bavelier mentioned that studies of child action gamers show that although spatial attention is matured by seven years of age, playing shooter games can change its developmental time course. Of the three types of top–down attention tested in children, each had a different maturational curve, but all were changed by action game play. Bavelier noted that given how ubiquitous video game play is among children in our society, what is called an experiment of nature is unfolding in front of our eyes, with possible repercussions on society that are currently poorly understand.

One of the techniques that Bavelier's lab uses to better understand the source of attentional enhancement after action game play is called *steady-state evoked potentials* and consists of presenting subjects with four different streams of visual information and measuring patterns of activation.[28] By flickering the information at different rates, each stream induces a unique pattern of activation over the participant's scalp that allows one to measure, and follow, the fate not only of an attended stream of information but also of unattended, potentially distracting information streams. Using a similar technique, Srinivasan and colleagues also concluded that action gamers excel at divided attention and actively suppress distraction.[28] In contrast, role game players efficiently enhance attended information but show no other changes. This and many other studies suggest that action gamers are not simply trigger-happy. Rather, they are dealing with incoming information more efficiently, whether visual or auditory. The most recent research that Bavelier described explores the possibility that by enhancing top–down attentional control, action video game play also fosters the ability to learn more efficiently.[24]

A key challenge for future research will be to determine the particular elements of action video games that are necessary for these effects. Bavelier believes there is a need for much more research, as the impact of the technology currently used throughout society is quite unintuitive. A recent study at Stanford University shows that undergraduates who report multitasking between many different media have very poor attentional control when measured in the lab; this is the case, despite these individuals being convinced they excelled at the laboratory tasks they were just evaluated on![29] In contrast, based on what Bavelier reported, playing action video games enhances attentional control, despite prompting an initial impression of being a mindless activity.

Bavelier closed by pointing out that playing action video games may thus be viewed as a tool for leveraging brain plasticity in various patient populations, be it in amblyopia or in reversing the cognitive consequences of aging. Playing video games is already under consideration as a training device/task for specific kinds of work. For example, laparoscopic surgeons that play action video games are better at performing surgery than colleagues who do not play them; they are more accurate and make fewer errors than colleagues without video play but with more experience in the surgery itself.[30] Pilots or other military personnel may similarly benefit from such video game play. The application of this research

to educational goals is also being considered; primary school children could use video game play to develop core abilities, such as number sense, object manipulation, or basic physics. More accurate representations within these systems could translate into better school performance.

Development of the neocortex

Krubitzer gave the fourth overview lecture in which she explored how cortical phenotypes develop across lifetimes, and how it changes within them.[31] Krubitzer's lab considers the common features of brain organization in many different species due to homology. She also demonstrates that similar types of modifications have been made to the neocortex in a variety of different mammals, suggesting that there are significant constraints on how evolution builds the neocortex. Throughout evolution, brains change in highly predictable ways, and Krubitzer's lab works to determine the factors that lead to specific phenotypic characteristics. One approach to determining which factors specifically contribute to phenotypic variability is to induce the types of changes to the developing brain that are thought to be contributing to evolutionary changes, and then examine the resulting brain organization. These types of studies allow one to postulate how transitions in phenotype occur.

Of course both genes and the environment play significant roles in generating the changes observed in different mammalian brains. Other major factors that contribute to phenotypic differences across mammals are the morphological and sensory specializations of an animal and the species-specific behaviors associated with specialized body parts.[32] This specialization leads to an enlargement of sensory domain allocation (the amount of cortex devoted to processing inputs from a particular sensory system) as well as cortical magnification (an enlargement of the representation of the specialized morphology within a cortical field). Consider the duck-billed platypus (*Ornithorhynchus anatinus*), a semiaquatic mammal endemic to eastern Australia, as a different, and more extreme, example of morphological/behavioral specialization. When it interacts with the world, it closes its eyes, ears, and nose. Thus, the only sensory inputs relaying information about the environment are coming from touch and electrosensory receptors on its bill, an extreme magnification of which exists within its primary sensory cortex. In fact, most of its neocortex in general is taken over by inputs from its bill, roughly 75%.[32]

These observations indicate that remarkable changes in the neocortex can be effected by altering peripheral morphology, sensory inputs, and the types of behavior associated with a given sensory receptor system. The effects of altering peripheral morphology and sensory-driven activity on the cortical phenotype have been assessed in the South American opossum.[33] This approach consists of removing all of their visual input and examining the effect of this loss on the functional organization and connectivity of their neocortex. There is no change in the size of their cortical sheet. However, what would normally be primary visual cortex (identified architectonically with myelin stains) becomes very small, but does not disappear. This is similar to what is observed in blind mole rats, which have very small eyes, covered by skin. They use their visual system only for setting their biological clocks, not for navigating the environment. Thus, with the loss of visual inputs, sensory systems associated with the lost system contract but do not disappear.

This complete loss of visual inputs in opossums also causes dramatic changes in the functional organization of the neocortex; all of what would normally be visual cortex is taken over by the auditory and somatosensory systems, and neurons in this area respond to somatosensory and auditory stimulation. Further, this cortex that would normally develop into visual cortex now receives connections from subcortical and cortical areas associated with processing auditory and somatosensory inputs.

The question is whether it is possible to direct this cross-modal plasticity following early loss of vision. The Krubitzer laboratory is beginning to answer this question by allowing opossums to develop in an enriched tactile environment following complete loss of vision, to determine if cortical plasticity can be directed and enhanced based on the sensory environment in which the individual develops. Will neurons be tuned to the enhancing stimulus? Will connections be modified such that inputs from somatosensory structures come to dominate the reorganized visual cortex? Can tactile discrimination be enhanced? In addition to examining the functional and anatomical differences that emerge with enhancement following sensory loss, the cellular composition of the reorganized cortex will be examined

using the isotropic fractionator method,[34,35] which involves homogenization of tissue, leaving cellular nuclei intact. Differential staining of these nuclei with DAPI (labels all nuclei) and NeuN (labels neurons) allows one to quantify whether neuronal versus non-neuronal cell numbers have changed and answer the question: Will there be more neurons in the enhanced brain and/or a greater density of neurons?

The role of early sensory experience in shaping cortical organization has been underscored by comparisons between the laboratory rat and the same species of wild-caught rat, meaning one that lives in a laboratory cage and one that lives out in the natural world.[35] Measurable differences have been found in the size of their auditory and primary somatosensory cortex and in the neuronal composition of the primary visual area. Laboratory rats have a larger auditory and somatosensory cortex compared to wild-caught rats (senses that were probably the least impacted by laboratory rearing). However, wild-caught rats have a greater density of neurons in their primary visual cortex.

These results have led the Krubitzer laboratory to examine the effects of natural differences in social rearing in voles.[36] Parenting styles of voles can be naturally divided into parents that have a lot of tactile contact with their young (high contact parents) and those that have significantly less tactile contact with their young (low contact parents). This contact occurs around the perioral facial area of voles (i.e., around the mouth and nose). Voles reared by high contact parents have a greater amount of primary somatosensory cortex devoted to processing inputs from the perioral facial region than those reared by low contact parents. In addition, there are differences in cortical connections between offspring reared by high versus low contact parents.

Genes also contribute to the cortical phenotype and impact cortical sheet size, cortical field size, cortical connections, and peripheral morphology. In addition, cellular mechanisms involved in plasticity can be genetically mediated. More complicated environmental factors, such as social learning and culture, also have a large impact and represent complex patterns of interacting sensory stimuli that impinge on the developing brain, generate changes in cortical organization and connectivity, and ultimately influence subsequent behavior. In all, brains develop to match the sensory context in which they

Ann. N.Y. Acad. Sci. 1292 (2013) 1–20 © 2013 The Authors. *Annals of the New York Academy of Sciences* published by Wiley Periodicals Inc. on behalf of The New York Academy of Sciences.

develop and generate appropriate, context-specific behaviors.

Development of educational ability

Nora S. Newcombe (Temple University) gave the fifth and final overview lecture, entitled "Play and Educational Outcomes." Newcombe discussed how early spatial learning is important in subsequent entry into STEM fields (i.e., science, technology, engineering, and mathematics) and how spatial abilities can be improved. Newcombe reviewed evidence from Wai, Lubinski, and Benbow[37] on the abilities of high school students and predictions of their occupational interests and paths. Factoring out verbal and mathematical ability and other background factors, the likelihood that people pursue certain occupations has been shown to relate to their spatial ability. Students who subsequently enter STEM fields have higher spatial skills.

In collaboration with David Uttal, Newcombe[38] performed a meta-analysis of training effects for five different classes of spatial skill. *Disembedding*, the ability to look for some specific pattern and pull it out, has the smallest effect size, and *spatial perception* (i.e., discern horizontal and vertical, with respect to gravity) has the largest. However, all of the effect sizes from their analysis are large enough to have practical importance. Their analysis leads to the following question: How high, in terms of spatial ability, do people have to be to major in STEM disciplines? Assuming the threshold for entering STEM disciplines remains the same, shifting the distribution with training may increase the proportion that can enter these disciplines. Additional work is needed to establish whether improvements in spatial ability actually have this effect when examined in rigorous randomized control studies.

One approach taken by Newcombe has been to compare men and women with high and low spatial ability who did or did not play *Tetris* over the course of a semester. Both women and men with high spatial ability improve more rapidly initially than they do later, although both continue to improve and do not generally reach a ceiling. Women with low ability initially improve more slowly than they do later. Comparing two groups that are equal on pretest assessments shows that the training group beats the other group on posttest assessments and on retests. Newcombe and colleagues have shown that these effects are durable.

Newcombe pointed out that according to the Wai *et al.*[37] study, teachers in K–12 tend to have lower spatial ability while they are in high school. In other words, students who will become K–12 teachers tend to have lower spatial ability than their peers. This presents a problem: How do we get teachers to develop scientists? Newcombe described five activities that students can do that can help develop spatial abilities, including spatial books and poems, puzzle play, paper folding, block play, and shape sorters.

Spatial language is very important for nurturing spatial development. Istvan Banyai's *Zoom* is an example of a book that can facilitate spatial learning by applying and exchanging spatial language. There are no words in the book and parents have to talk about what they are seeing in different scenes. The scenes are shown from different perspectives and unfold at different scales, so there are many opportunities to use spatial language to continuously explain what is going on.[39]

Jigsaw puzzle play can also foster spatial development, something that may differ by gender; particularly between two and four years of age, gender differences can be seen. The quality of puzzle play has been found to be higher in boys. For example, they can do more difficult puzzles, their parents are more engaged, and they use more spatial language. However, the quality of the puzzle has a greater impact on girls. Whereas boys perform higher regardless of puzzle quality, girls perform better if they are given harder puzzles. This hints at a complex interaction where what children bring to the interaction is important.[40]

Play with blocks is also important. Blocks can be arranged in different ways, which allows for comparisons between free play, play with prebuilt structures, and guided play. In free play, the blocks are just there to play with, whereas in guided play, there is a defined structure to work towards and an assembly diagram. Results suggest that if parents interact during play with blocks at all, they will use more spatial language, but they increase spatial language even more in the guided play condition. In other words, the mere presence of blocks increases the use of spatial language, and in correlational and longitudinal studies, spatial language is associated with higher spatial skill.[41]

Newcombe, Kathy Hirsh-Pasek (Temple University), and graduate student Justin Harris have developed an assessment of spatial folding that works for young children. Their results show that children do not usually perform above chance when they are younger than five years of age. However, engaging in this kind of activity more systematically might enhance spatial skill. Shape sorters are another important tool for improving and assessing spatial ability. Work with shape sorters shows that children often see only typical shapes; for example, they do not see a hole in a triangle. Typically, kids are shown only equilateral triangles; the result is they can think that all triangles are like these. A survey of the kinds of triangles that children are shown in math books up to the age of four years shows that they are all conventional triangles (i.e., equilateral triangles). In comparing guided play, didactic instruction, and enriched free play, Kelly Fisher has shown that guided play does the best in getting them to learn the correct definition of shapes.[42,43] Catherine Tamis-LeMonda (New York University) commented that a combined effort by a child and a parent would allow a child to quickly reach the right answer, but that the child would not then show significant transfer.

Bavelier pointed out that people improve on different spatial skills with 10–20 hours of practice. She mentioned that there is other work on expert *Tetris* players suggesting that they excel at rotating *Tetris*-like shapes but not other shapes. *Tetris* experts may not engage any longer in the effortful process of mental rotation, but rather use lookup tables, having learned the mapping of shapes to board. This pattern of behavior suggests that transfer may be a U-shaped function. During the early phases of acquiring an effortful task, the need for attention and executive control may train such domain-general resources and benefit others, relatively different tasks allowing broad transfer. Yet when expertise develops, the complexities of the task are learned at a more procedural level, releasing effortful processing and enabling expert performance. The price to pay is that the knowledge is now much more specific, implying only limited transfer.

Newcombe replied to Bavelier by pointing out that there are a limited number of shapes in *Tetris*, and that there are different routes for performance improvement. One approach is to memorize all the shapes in all possible positions, which has a particular signature in ERP. This is the route that most people would take to becoming expert *Tetris* players, but there may still be some generalized transfer

of mental rotation skills. The general idea is that the richness of the environment within a game is important and may determine the amount of transfer that can result from practice. Tamis-LeMonda pointed out that speed of learning and transfer can be in opposition.

Ghajar asked about time constraints in *Tetris* and in shooter games. He asked whether there are differential effects of play for engaging spatial ability within limited time constraints. Is there a separation between learning how to mentally rotate things and doing them in a very short time frame? Ghajar also asked about what is transferring. Do players anticipate better and show reductions in reaction time or do they do something else? Newcombe replied that the only related work that she is aware of related to these questions is on gender differences in mental rotation. The results of this work suggests that if timing pressure is eliminated, women are equal to men. But that turns out not to be true according to other studies, implying that there is more going on than just timing. Adolph added that infant control and anticipation are some the most important issues researchers are working on.

Bavelier pointed out that action video games have a rich temporal structure. Players have goals at many different time scales and are effective at taking all of them into consideration. Many action video games are built to allow the player a good scaffolding of knowledge and, at the same time, always keep the players in a world where they are challenged. In order to have a good game, it is important for players not to be able to develop a routine by which they can know what will happen next. Also, the video game industry has figured out how to build games with such a delicate balance, a feature that may make them not only maximally interesting, but also good learning tools. The notion of just-right challenges in the field of learning is not new, but it is very hard to quantify, and is clearly individual-dependent, making its implementation quite delicate.

Scott Eberle (Strong National Museum of Play) closed the discussion on time and anticipation by pointing out that there are other variables to consider, and that changing them may provide some insight into what it means to be playful. Chess players sometimes turn their back on the board, relying on their memory of piece position. Puzzle players sometimes turn a puzzle upside down so they cannot see the picture, relying on shape alone to fit pieces together. It may be that altering a game in a way that is novel and surprising is what it means to be playful. Playfulness, therefore, may be a way of expanding expertise to new abilities.

Gopnik connected things back to the definition of play. There seems to be two different dimensions that are often conflated: (1) play is something that children are engaged in independently or is something that involves others and has a didactic or pedagogic component; and (2) play is designed to accomplish a particular goal or is broad ranging and exploratory. These are very different dimensions. Results suggesting that kids do better in guided play than in free play may in part be a consequence of tapping into something where developing a specific skill is what is best for an individual; but one could arrive at a different result if doing something broader was beneficial and more specificity was limiting.

Working group sessions

In the working group sessions, the participants were assigned to one of four groups, and three topical questions were provided to them by the workshop organizers. Following a discussion period, each group's spokesperson provided a summary and opened discussion with all workshop participants. The posed questions and selected points from the discussions are provided below.

Question 1: What current research evidence is there to connect play activities and the development of attention to the ability to learn in formal settings, like the classroom?

Dima Amso's (Brown University) group added to earlier discussions about the definition of play and which aspects of attention might be important for classroom learning. Play is an active and emergent process of interaction with the outside world. It encompasses exploratory processes, which subsequently give rise to either serendipitous or intentional discovery of something that may be important. Additional bouts of exploration within play may follow in a cascade, as one comes into contact with the world and discovers additional things to work on. There is a balance between mastery and exploration, with novelty engaging attention. Motivation, reward, and pleasure are also likely to be involved, because they all foster play and may support sustained attention on one task. In other

words, these elements may add drive when working out a problem. Variability and repetition may also be particularly important for fostering play that will develop attention.

Specific kinds of play may connect to specific kinds of attention. Pretend play may, for example, facilitate the ability to consider someone else's perspective and to switch between different tasks, whereas puzzle play may facilitate sustained attention to a difficult assembly task. Similarly, different aspects of attention may relate to different activities in classroom learning. Sustained attention may, for example, connect to listening to the teacher throughout a lesson. Task switching that is facilitated by play would allow for working on one problem set and then another without a big drop in performance. Similarly, planning ability from play would allow for such things as picking up particular objects while in the process of assembling a model toy. A related observation is that planning becomes more abstract as children develop. They begin to use rules to plan their play, and this transition would seem to, at least in part, relate to the development of the frontal system.

Other evidence on the connection between play, attention, and learning comes from comparisons of curricula and outcomes of Montessori education with those of the Tools of the Mind curriculum. Montessori centers on self-construction by means of interaction with the environment and on the idea of an innate path of psychological development. Montessori programs allow children to make their own choices within structured environments. Montessori emphasizes keeping activities challenging. Once a task becomes easy, the child goes on to the next level. This is repeated until they master a skill, at which point they move on to something else. Although Montessori may undervalue social interaction, it facilitates the development of attention even into adolescence. Tools of The Mind centers on the development of self-regulation (i.e., executive function). Executive function can be defined as the ability to regulate social, emotional, and cognitive behaviors. Certain interventions with young children promote executive function, which in turn correlates with children's achievement in literacy and mathematics. In comparison to Montessori, children have much more time for pretend play, which may be particularly important for developing executive function.

Bavelier's group added another definition of play: activities that are fun, voluntary, and flexible. Such activities involve active engagement, the absence of extrinsic goals, an element of pretend, and are usually done in an environment where there are very few consequences. Bavelier's group pointed out that there is very limited evidence connecting play to cognitive ability, including attention, if the above definition of play is applied. However, evidence suggests that children who engage in free play have better self-control, an observation that connects more to Montessori than to Tools of The Mind. The latter is more contained and takes place in a restricted environment. In light of this point, Bavelier's group also pointed out that it would be helpful to know what would happen if children played in an environment designed mostly by themselves.

Bavelier's group continued by discussing other observations that suggest that if children engage in an activity frequently during free play, they will get better at it. Play that involves a lot of language interactions will, for example, make children all the more ready for language tasks. Play that involves organizing objects by number or by manipulation into different groups will, for example, build a better sense of numbers and numeracy. A similar link among attention and executive control does not seem to have been developed, although related work on discovery-based learning suggests that it can be important to ensure that children understand the underlying conceptual framework of a problem rather than just knowing how to solve it. This approach, with an emphasis on concepts, takes a long time, which is consistent with a trade-off between knowledge acquired and time spent on a task. A final point is that intrinsic and extrinsic motivation could be important factors. There are cultures in which children are not allowed to play by themselves in that all aspects of play are directed.

Julie A. Fiez's (University of Pittsburgh) group introduced a third definition of play, one that takes a more comparative species approach and which can be described in terms of specific criteria: activity that is not immediately functional; is pleasurable; occurs in a relaxed field; is repetitive, but not stereotyped; and is spontaneous in nature. Each of these criteria can be evaluated across different species and along a developmental trajectory within one species. This definition prompts a number of issues, includ-

ing the kinds of situations or activities that increase the likelihood of play and whether they encourage the exploration of variability and the causal mechanisms for initiating play. Species that use learning as one of their core survival mechanisms tend to have more extended play over the course of their development. Play allows for the opportunity to learn to make predictions and to reason about variability and causality in the world. In addition, in linking certain types of play to certain types of improvements, play that involves a lot of social interaction will lead to improvements in social interaction, whereas play that includes a focus on causality will show transfer to other processes involving reasoning about causality.

Newcombe's group pointed out that there are many different kinds of play (e.g., puzzle play, swinging on a swing, pretending to be a fireman) that may have effects on attention. In fact, each type of play would seem likely to have an impact on what follows in substantive ways. Yet, importantly, little is known about the transfer problem in most cases. For example, it is not known, even from correlational analyses, whether children who are more likely to become firemen as adults spend their playtime in activities such as bouncing around playing a drum. Also, a distinction should be made between extreme environments and normal variation.

Suzanne Gaskins (Northeastern Illinois University) pointed out that attention could be defined differently than it usually is within the cognitive neuroscience community. Instead of the three subtypes measured in the ANT, attention could be considered to be something one does to survive in the street of a poor neighborhood. This alternative notion of attention connects to well-known work by Walter Mischel, Yuichi Shoda, and Monica Rodriguez at Columbia University on delay of gratification. It may be that for those people who develop in a low socioeconomic status or a war zone, it would make sense for them to "take the cookie" (i.e., not to delay gratification). To the extent that an educational process emphasizes, or draws upon, certain abilities that may be practiced within play, play may confer benefits. Yet evidence for this on play in the classroom is not available. Questions to be explored include whether children learn better from play than from direct instruction, and how structured play relates to learning.

Question 2: What is the role of timing (i.e., rhythm and cadence) in play activities, and how might this influence the development of attention and learning?

Amso's group pointed out that rhythms that are in the world include repetition, something that may be important for learning. Repetition provides a lot of information that may obviate the need for sophisticated attention systems early in life. It may be significant that the repetition that is exploited in children's play can span multiple modalities. For example, patty-cake, the game in which two players clap hands while singing an English nursery rhyme, involves synchrony across motor, auditory, tactile, and proprioceptive inputs. This multimodal source of repeated information may be particularly important for learning.

Input following a regular temporal pattern may allow children to build up a structure for processing incoming information. Pervasive and persistent temporal patterns allow for predictions, which in turn allow for the creation of error signals, a key element of learning. That which is external may become internalized, and may help make predictions about what is going on. This can be thought of in terms of cascades of "I know this is coming up and then it did." Related work by Michael Goldstein and colleagues at Cornell University emphasizes the role that social interactions play in learning and timing. If a mom has headphones on and cannot hear what her baby is saying but still responds at the right times, the baby will still learn, thus emphasizing the significant role that timing and cadence play in the early environment. An open question is whether this sort of timing requires motor patterns.

Bavelier's group agreed that timing is important in play and that it helps to structure it. For example, play allows for social interactions to become highly structured. As above, timing makes play more rewarding by facilitating the creation of scaffolding for performing cycles of prediction, error prediction, and error measurement. In other words, it allows for the collection of feedback. But timing is not the only thing that can be used for structuring play; spatial information plays a significant role as well. Bavelier's group pointed out that related work suggests that the dopamine pathway is important for controlling reward-based learning and decision making. An interesting element from this work is

that the dopamine pathway is important not just for getting rewarded for getting something right but for intrinsic reward as well. Predicting which reward is expected releases dopamine at the time of the prediction, not at the point of experiencing the actual rewarding event.

Although the link between timing in play and learning is, in some cases, clear in the opinion of Bavelier's group, the link between timing in play and attention is not. Studies on training with an interactive metronome, for example, do not make a connection to attention. Similarly, although there is an important push in social interactions for synchronicity (e.g., in joint attention), it is not known whether such synchronicity actually leads to greater joint attention. To this point, there is some evidence that interactions that seek to draw attention together are more common in cultures where mothers pay less attention than in cultures where they pay more attention—in other words, in cultures where children are used to receiving attention. And although it is unclear how this translates to effects on different aspects of attention and learning, it suggests that context is going to be important and is paying off in the timing of the interaction.

Fiez's group asked whether timing is particularly special for play or whether it is just a pervasive element for species living in an environment. Some consensus went toward the latter, but it was acknowledged that play that centers on timing and cadence does tend to be particularly attractive and may be especially engaging. Language play, for example, may be likely to involve play where timing is important, although this type of reasoning starts to become teleological, blurring what is causing what. A similar idea was discussed on the connection of play to learning and attention. Skills that children practice through play will lead to improvements in those skills, but that is not a special benefit only for timing. The basal ganglia system stands at the intersection of timing prediction and reward or pleasure signals. To the degree that this system supports a biological clock, individual differences in timing and in the ability to synchronize timing may be important in maintaining interindividual play interactions. From research in rats and playground behavior of children, it seems plausible that this could be an important mechanism for generating productive play experiences.

Newcombe's group pointed out that timing is obviously important for play activities and the development of attention and learning; almost everything has a temporal component: neuronal activity changes in time; rhythms are known to help memory and guide the temporal allocation of attention; rhythmic movements begin at the fetal stage; and contingent timing helps attribute agency to a partner.[44,45] McAuley added that understanding the role of timing and cadence in play, and the consequences for development, requires consideration of the range of rates (tempos) in which a child is able to perceive cadence and track events in time. Notably, we live in a particular temporal world where if successive events are too separated in time, they are perceived as isolated events; for adults, this temporal integration window is about 2–3 s, whereas for children, this window seems to be closer to 1 second.

Moreover, within this temporal integration window, children are generally tuned in to faster tempos than are adults, with preferred tempo slowing across the life span. The implication for play and attention here is that if a child has a narrower range of tempos within which she can connect two events in time and a faster preferred tempo, this places developmental constraints on how well a child will be able to track events in time. In this regard, rhythmic play activities may serve an important function because they have the potential to entrain attention to the time scale of the engaged play. Hierarchically structured play activities, including those involving music, have the added advantage that they may help bootstrap the development of attention to increasingly longer time spans.

Discussion of this question closed by connecting to a different question: Is there a reason to think that an activity such as rhythmic training (with music training as a specific type) could be helpful in developing attention or in learning? Could such training expand the trainee's temporal window of integration (i.e., the time scale with which they connect and track events)? Work on these questions suggests that music training can slow preferred tempos,[46] which is to say the speed at which information processing is optimal changes and becomes more matured. Related work by Nina Kraus and colleagues at Northwestern University suggests that musical training can impact learning; and work at the Temporal Dynamics of Learning Center at the University of California, San Diego suggests that training in Gamelan drumming can impact attentional ability.

Question 3: How can we apply what we currently know about the relationships between play, attention, and learning to better design early interventions for children with attention and learning disabilities?

Amso's group suggested that, assuming that there is sufficient evidence linking play to development, one approach to answering this question would be to look at different trajectories following different kinds or amounts of play. One advantage to using play as a means of improving particular skills is that it is fun and often more engaging than straightforward practice. Tamis-LeMonda added that even if research confirms the importance of play, bringing it to public policy is going to be incredibly difficult: convincing schools or therapeutic teams that they should switch from practice to play would be difficult and something the field would need to tackle.

Bavelier's group described a different approach to this question, which is to consider a specific disorder (e.g., autism), and ask what is known about the effects of play interventions. Of relevance to this approach are experiments using the game *Second Life*, which requires identifying with avatars and interacting with others. A player identifies with an avatar that he creates from scratch and hence can be whatever he wants (e.g., a little boy can become a woman with long eyelashes). One idea proposed putting autistic people "on their own island" and recording their interactions (i.e., how much they are looking at each other); in fact, over a period of six months they become more social. Bavelier's group added that other results suggest that autism is not a problem of attention and not related to play in any way, but instead, that it may involve a lack of maturation of the gamma-aminobutyric acid (GABA) receptor during critical developmental periods. As a consequence, the GABA receptor ceases to be excitatory, which may be associated with the development of autism. There are intriguing studies supporting this in which antidiuretics are given to autistic children; these drugs act on the GABA cascade and have been argued to confer significant improvements within a few days of taking them. Considering another disorder, dyslexic children suffer from phonological problems, but they also seem to have attention-related problems. Video games have been used to retrain attention of dyslexic children. The approach to this research is to first work on attention and the child's phonological learning.

Bavelier's group pointed out that another related area is the field of play therapy, which would benefit from carefully controlled assessments. Work in this area includes studies in which children who have difficulty focusing and do not enjoy reading are supported by a reader robot. The robot asks open-ended questions, which soothes the children and facilitates learning. Rigorous studies based on these kinds of approaches, however, have not yet been done. Fiez's group added that more basic scientific research to address these issues would be informative, but also that such science should move toward building bridges between them. Specific funding that focuses on bridging these issues could be helpful, but there is some doubt as to whether, at this stage, a given funding review board would be able to differentiate among the issues to determine what is informative and whether it is likely to bear fruit. Fiez's group commented that in discussing intervention, their impression was that a lot of intervention work is not conducted rigorously and may be prone to the Hawthorne effect in which observed effects are not a consequence of changes in a group's behavior, but are actually related to the social situation of the experiment and the treatment the group receives.

Despite these limitations, there are good clues that could guide the choice of interventions that are likely to be effective. Focus should be placed on the zone of proximal development, the space between what learners can do with assistance and what they cannot do, and on play that taps into content or ability that is desired. Structured play activities may be particularly likely to yield benefits that generalize and are worth exploring further. Discussion on these issues also included consideration of a different approach where children are simply provided with more opportunities for play. For example, in the Finnish educational system, children have more time to play, an approach that has been applied in the corporate world and for adults at companies like Google, which provide less structured time in order to promote innovation and efficiency.

Newcombe's group mentioned work by Cole Galloway at the University of Delaware on children diagnosed with cerebral palsy (CP) who benefit from using scooters to move around, which has wide-ranging cognitive effects. This highlights the importance of understanding the nature of any particular disability and its consequences on normal development. Discussion also included ADHD and the consensus that the specific causes of ADHD are not yet fully understood, and that it seems unlikely that undifferentiated play would provide much help. Children with ADHD may actually need more structured environments than those without ADHD. Another issue to consider is that cultures or institutions can have different definitions of attention or different ideas of how attention manifests, leading to differences in the number of children who are classified as atypical. A final point is that a large longitudinal study should provide greater insights into how play relates to learning and attention.

Conclusions

The participants of this workshop considered many aspects of play, essentially different compositions of what play accomplishes. Play is an active and emergent process of engagement with the world, which encompasses exploratory processes. It is repetitive, but not stereotyped, and is spontaneous in nature. Along with play there are transitions in body size and sensorimotor development, each of which could facilitate a reciprocal developmental process during play.

It may also be that certain elements of play generalize, whereas others do not. Evidence was reviewed that shows that improvements in playing certain games generalize, probably by sharpening top–down attention. Longitudinal studies that look at trajectories following different kinds or amounts of play could provide crucial insights on many of these issues. A first question might be to explore what happens if children cannot play, or cannot play very much. Observing children in their natural environment and developing ways to quantify play behavior would be key to knowing how play varies across contexts.

Timing in play was considered in relation to the development of attention and the ability for subsequent learning. Dynamic temporal patterns, such as cadence in speech, could be important for learning and repetition. Timing is such a pervasive function that teasing out its contribution could be difficult. Certainly, performance improvement with practice involves transitioning from reactive to predictive timing, so different kinds of timing need to be considered.

There was some agreement that play could be used as an intervention in severe learning disorders. Playing certain immersive reality games has improved the social abilities of young people diagnosed with autism, who have reduced social play and impaired predictive timing abilities. There was also some agreement that unstructured play may not be beneficial for children diagnosed with ADHD. While some evidence exists for the utility of play intervention therapy in pathological conditions, there is little work describing the specific play dynamics that were necessary for therapeutic efficacy.

Questions for future work

The workshop identified a number of important questions, which could be the focus of upcoming research:

The role of play in development

- How is the development of attention and learning influenced by play, and by structured and unstructured play in particular?
- How are the development of the cerebellum (e.g., granule cell migration and synaptogenesis) and the formation of cerebellar–cortical connections influenced by play?
- How does play that is composed of particular combinations of activities relate to the development of a particular combination of abilities later in life?
- What are appropriate metrics for assessing attention and learning in young children?
- In what cases would children learn better from play than from direct instruction?

Variation in play

- How does play vary across cultures?
- Which elements of video games are important for improving attention and, if possible, generalizing to other cognitive functions?
- What are the effects of playing video games on attention and social skills?
- What is the role of play in an evolutionary context?

Play and timing

- How does play facilitate a transition from reactive to predictive sensory processing?

- How do neural networks that support anticipatory timing (i.e., those that underlie contingent negative variation) develop?
- What are the differential contributions of spatial and temporal regularities in structuring play?
- Does movement need to be connected to temporal elements in play in order to drive any effects on learning and on the formation of internal temporal representations?

Play as an intervention

- How can the abilities that children typically develop through play be modified to facilitate improved learning in more formal environments?
- Would children with attention or learning disorders benefit from play-based interventions?
- Would rhythmic training facilitate improved attention and learning abilities?

Acknowledgements

We thank Kurt Fischer and Richard Ivry for bringing the workshop "Play, Attention, and Learning: How Do Play and Timing Shape the Development of Attention and Facilitate Classroom Learning?" together as members of the steering committee; the workshop participants Neil Albert, Alison Gopnik, Marc Bornstein, Stuart Brown, Scott Eberle, Kurt Fischer, Suzanne Gaskins, Asif Ghazanfar, Usha Goswami, Kathy Hirsh-Pasek, David Kanter, Alice Luo Clayton, Susan Magsamen, Bruce McCandliss, Brian Rakitin, Philippe Rochat, Stephen Siviy, Catherine Tamis-LeMonda, and Martin Wiener; Brooke Grindlinger, Sonya Dougal, and Melinda Miller of the New York Academy of Sciences for developing, participating in, and conducting the workshop; and Douglas Braaten, editor-in-chief of *Annals of the New York Academy of Sciences*, for helping to prepare this summary. The conference was presented by the New York Academy of Sciences and the Brain Trauma Foundation and supported by educational grants from the Leon Levy Foundation and NewYork-Presbyterian Hospital.

Conflicts of interest

The authors declare no conflicts of interest.

References

1. Ghajar, J. & R.B. Ivry. 2009. The predictive brain state: asynchrony in disorders of attention? *The Neuroscientist* **15:** 232–242.

2. Franchak, J.M., K.S. Kretch, K.C. Soska & K.E. Adolph. 2011. Head-mounted eye tracking: a new method to describe infant looking. *Child Development* **82:** 1738–1750.

3. Franchak, J.M., K.S. Kretch, K.C. Soska, J.S. Babcock & K.E. Adolph. 2010. Head-mounted eye-tracking of infants' natural interactions: A new method. In *Symposium on Eye Tracking Research and Applications, Austin, TX.*

4. Adolph, K.E., W.G. Cole, M. Komati, J.S. Garciaguirre, D. Badaly, J.M. Lingeman, G. Chan & R.B. Sotsky. 2012. How Do You Learn to Walk? Thousands of Steps and Dozens of Falls per Day. *Psychological Science* **23:** 1387–1394.

5. Karasik, L.B., C.S. Tamis-LeMonda & K.E. Adolph. 2011. Transition from Crawling to Walking and Infants' Actions with Objects and People. *Child Development* **82:** 1199–1209.

6. Kretch, K.S., J.M. Franchak, J.L. Brothers & K.E. Adolph. 2012. Effects of locomotor posture on infants' visual experiences. Conference presentation at the XVIII Biennial International Conference on Infant Studies, Minneapolis, MN.

7. Karasik, L.B., K.E. Adolph, C.S. Tamis-LeMonda & A.L. Zuckerman. 2012. Carry on: spontaneous object carrying in 13-month-old crawling and walking infants. *Developmental Psychology* **48:** 389–397.

8. Karasik, L.B., E.C. Celano, C.S. Tamis-LeMonda & K.E. Adolph. 2012. Maternal response to infant object sharing. Conference presentation at the XVIII Biennial International Conference on Infant Studies, Minneapolis, MN.

9. Soska, K.C., K.E. Adolph & S.P. Johnson. 2010. Systems in Development: Motor Skill Acquisition Facilitates 3D Object Completion. *Developmental Psychology* **46:** 129–138.

10. Posner, M.I., S.E. Petersen, P.T. Fox & M.E. Raichle. 1995. Localization of in the Cognitive Human Brain. *Science* **240:** 1627–1631.

11. Petersen, S.E. & M.I. Posner. 2012. The Attention System of the Human Brain: 20 Years After. *Annu Rev Neurosci* **35:** 73–89.

12. Fan, J., B.D. McCandliss, T. Sommer, A. Raz & M.I. Posner. 2002. Testing the efficiency and independence of attentional networks. *Journal of Cognitive Neuroscience* **14:** 340–347.

13. Eriksen, B.a. & C.W. Eriksen. 1974. Effects of noise letters upon the identification of a target letter in a nonsearch task. *Perception & Psychophysics* **16:** 143–149.

14. Rueda, M.R., J. Fan, B.D. McCandliss, J.D. Halparin, D.B. Gruber, L.P. Lercari & M.I. Posner. 2004. Development of attentional networks in childhood. *Neuropsychologia* **42:** 1029–1040.

15. Fan, J., J.I. Flombaum, B.D. McCandliss, K.M. Thomas & M.I. Posner. 2003. Cognitive and Brain Consequences of Conflict. *NeuroImage* **18:** 42–57.

16. Diamond, A., W.S. Barnett, J. Thomas & S. Munro. 2007. Preschool program improves cognitive control. *Science* **318:** 1387–1388.

17. Diamond, A. & K. Lee. 2011. Interventions shown to aid executive function development in children 4 to 12 years old. *Science* **333:** 959–964.

18. Curtis, C.E. & D. Lee. 2010. Beyond working memory: the role of persistent activity in decision making. *Trends in Cognitive Sciences* **14:** 216–222.

19. Walter, W., R. Cooper, V.J. Aldridge, W.C. McCallum & A.L. Winter. 1964. Contingent negative variation: an electric sign of sensori-motor association and expectancy in the human brain. *Nature* **203:** 380–384.

20. Halperin, J.M., V. Sharma, E. Greenblatt & S.T. Schwartz. 1991. Assessment of the continuous performance test: reliability and validity in a nonreferred sample. *Psychological Assessment* **3:** 603.

21. Eisenberger, N.I., M.D. Lieberman & K.D. Williams. 2003. Does rejection hurt? An FMRI study of social exclusion. *Science* **302:** 290–292.

22. Masten, C.L., N.I. Eisenberger, L.A. Borofsky, K. McNealy, J.H. Pfeifer, M. Dapretto, *et al.* 2011. Subgenual anterior cingulate responses to peer rejection: a marker of adolescents' risk for depression. *Development and Psychopathology* **23:** 283.

23. Bavelier, D., C.S. Green, D.H. Han, P.F. Renshaw, M.M. Merzenich & D.a. Gentile. 2011. Brains on video games. *Nature Reviews Neuroscience* **12:** 763–768.

24. Green, C.S., A. Pouget & D. Bavelier. 2010. Improved probabilistic inference as a general learning mechanism with action video games. *Current Biology* **20:** 1573–1579.

25. Li, R., U. Polat, W. Makous & D. Bavelier. 2009. Enhancing the contrast sensitivity function through action video game training. *Nature Neuroscience* **12:** 2008–2010.

26. Hubert-Wallander, B., C.S. Green & D. Bavelier. 2011. Stretching the limits of visual attention: the case of action video games. *Wiley Interdisciplinary Reviews: Cognitive Science* **2:** 222–230.

27. Bavelier, D., R.L. Achtman, M. Mani & J. Föcker. 2011. Neural bases of selective attention in action video game players. *Vision Research* **61:** 132–143.

28. Mishra, J., M. Zinni, D. Bavelier & S.a. Hillyard. 2011. Neural basis of superior performance of action videogame players in an attention-demanding task. *The Journal of Neuroscience* **31:** 992–998.

29. Ophir, E., C. Nass & A.D. Wagner. 2009. Cognitive control in media multitaskers. *Proceedings of the National Academy of Sciences of the United States of America U.S.A.* **106:** 15583–15587.

30. Rosser, J. C., P. J. Lynch, L. Cuddihy, D. a Gentile, J. Klonsky & R. Merrell. 2007. The impact of video games on training surgeons in the 21st century. *Archives of Surgery* **142:** 181–186; discusssion 186.

31. Krubitzer, L. 2009. In search of a unifying theory of complex brain evolution. *Ann. N.Y. Acad. Sci.* **1156:** 44–67.

32. Krubitzer, L. 2007. The magnificent compromise: cortical field evolution in mammals. *Neuron* **56:** 201–208.

33. Larsen, D. D. & L. Krubitzer. 2008. Genetic and epigenetic contributions to the cortical phenotype in mammals. *Brain Research Bulletin* **75:** 391–397.

34. Herculano-Houzel, S. & R. Lent. 2005. Isotropic fractionator: a simple, rapid method for the quantification of total cell and neuron numbers in the brain. *J. Neurosci.* **25:** 2518–2521.

35. Campi, K.L., C.E. Collins, W.D. Todd, J. Kaas & L. Krubitzer. 2011. Comparison of area 17 cellular composition

in laboratory and wild-caught rats including diurnal and nocturnal species. *Brain Behav. Evol.* **77:** 116–130.

36. Krubitzer, L., K.L. Campi & D.F. Cooke. 2011. All rodents are not the same: a modern synthesis of cortical organization. *Brain Behav. Evol.* **78:** 51–93.

37. Wai, J., D. Lubinski & C.P. Benbow. 2009. Spatial ability for STEM domains: Aligning over 50 years of cumulative psychological knowledge solidifies its importance. *Journal of Educational Psychology* **101:** 817–835.

38. Uttal, D.H., N.G. Meadow, E. Tipton, L.L. Hand, A.R. Alden, C. Warren & N.S. Newcombe. 2012. The Malleability of Spatial Skills: A Meta-Analysis of Training Studies. *Psychological bulletin.*

39. Szechter, L.E. & L.S. Liben. 2004. Parental guidance in preschoolers' understanding of spatial-graphic representations. *Child Development* **75:** 869–885.

40. Levine, S.C., K.R. Ratliff, J. Huttenlocher & J. Cannon. 2012. Early puzzle play: a predictor of preschoolers' spatial transformation skill. *Developmental Psychology* **48:** 530–542.

41. Ferrara, K., K. Hirsh-Pasek, N.S. Newcombe, R.M. Golinkoff & W.S. Lam. 2011. Block Talk: Spatial Language During Block Play. *Mind, Brain, and Education* **5:** 143–151.

42. Fisher, K.R., K. Hirsh-Pasek, N.S. Newcombe & R.M. Golinkoff. 2013. Taking Shape: Supporting Preschoolers' Acquisition of Geometric Knowledge Through Guided Play. *Child Development.*

43. Harris, J., N.S. Newcombe & K. Hirsh-Pasek. 2013. A New Twist on Studying the Development of Dynamic Spatial Transformations: Mental Paper Folding in Young Children. *Mind, Brain, and Education* **7:** 49–55.

44. Jones, M.R., H. Moynihan, N. MacKenzie & J. Puente. 2002. Temporal aspects of stimulus-driven attending in dynamic arrays. *Psychological science* **13:** 313–319.

45. Miller, J.E., L.A. Carlson & J.D. McAuley. 2013. When What You Hear Influences When You See Listening to an Auditory Rhythm Influences the Temporal Allocation of Visual Attention. *Psychological Science* **24:** 11–18.

46. Drake, C., M.R. Jones & C. Baruch. 2000. The development of rhythmic attending in auditory sequences: attunement, referent period, focal attending. *Cognition* **77:** 251–288.

Ann. N.Y. Acad. Sci. ISSN 0077-8923

New paradigms for treatment-resistant depression

Carlos Zarate,[1] Ronald S. Duman,[2] Guosong Liu,[3] Simone Sartori,[4] Jorge Quiroz,[5] and Harald Murck[6,7]

[1]Experimental Therapeutics & Pathophysiology Branch, Division of Intramural Research Program, National Institute of Mental Health, National Institutes of Health, and Department of Health and Human Services, Bethesda, Maryland. [2]Department of Psychiatry and Neurobiology, Yale University School of Medicine, New Haven, Connecticut. [3]Tsinghua-Peking Center for Life Sciences, School of Medicine, Tsinghua University, Beijing, China. [4]Department of Pharmacology and Toxicology, University of Innsbruck, Innsbruck, Austria. [5]Child and Adult Psychiatry, Neuroscience Translational Medicine, Roche, Nutley, New Jersey. [6]Covance, Princeton, New Jersey. [7]Psychiatric Clinic of the Philipps-University of Marburg, Marburg, Germany

Address for correspondence: Harald Murck, MD PhD, Senior Medical Director, Neuroscience Medical and Scientific Services, Covance, 206 Carnegie Center, Princeton, NJ 08540. Harald.Murck@Covance.com

Clinical depression is a serious mental disorder characterized by low mood, anhedonia, loss of interest in daily activities, and other symptoms, and is associated with severe consequences including suicide and increased risk of cardiovascular events. Depression affects nearly 15% of the population. The standard of care for the last 50 years has focused on monoamine neurotransmitters, including such treatments as selective serotonin reuptake inhibitors (SSRIs) and serotonin–norepinephrine reuptake inhibitors (SNRIs). However, these treatments have significant limitations: they can take weeks before showing mood-altering effects, and only one to two out of ten patients shows clinical effects beyond those associated with placebo. A major paradigm shift in research into the treatment of depression is underway, based on promising results with the glutamatergic NMDA receptor antagonist ketamine. Further research has demonstrated the significance of glutamatergic pathways in depression and the association of this system with the stress pathway and magnesium homeostasis. Treatment with NMDA receptor antagonists and magnesium have shown the ability to sprout new synaptic connections and reverse stress-induced neural changes, opening up promising new territory for the development of drugs to meet the unmet need in patients with clinical depression.

Keywords: depression; SSRI; SNRI; ketamine; glutamate antagonist; NMDA; scopolamine; magnesium; glutamine; CP-AMPA

Introduction

Research into the glutamatergic mechanism of depression is an important avenue to identify new treatments for depression. Several recent developments have come together to enable this. Fundamental findings in this area date back to the early 20th century. An early report on the mechanism of a compound, which has since been identified as an *N*-methyl-D-aspartate (NMDA) antagonist, was published in 1921 by Weston *et al.*[1] for the treatment of agitation, mostly in depressed patients. Much later, treatment with amantadine[2] demonstrated beneficial effects.[3] However, at that time glutamate—and hence the NMDA receptor—was not regarded as a neurotransmitter, and it was only accepted as such in the early 1980s.[4]

A clear theoretical foundation for the use of NMDA receptor antagonists in depression has been developed since 1990.[3] However, these early findings did not significantly influence antidepressant drug development until recently, as it had been almost exclusively informed by the monoamine hypothesis of depression.[5] In the early 21st century, the clinical limitations of the monoaminergic approach became apparent with the recognition of the relatively low efficacy of current treatment strategies.[6,7]

The serendipitous observation by Berman *et al.*[8] of the rapid response of ketamine in depression, later confirmed by Zarate,[9] led to the current focus in this area in both academia and industry. Extensive preclinical characterization of the effects of ketamine has partly illuminated its mechanism of

doi: 10.1111/nyas.12223

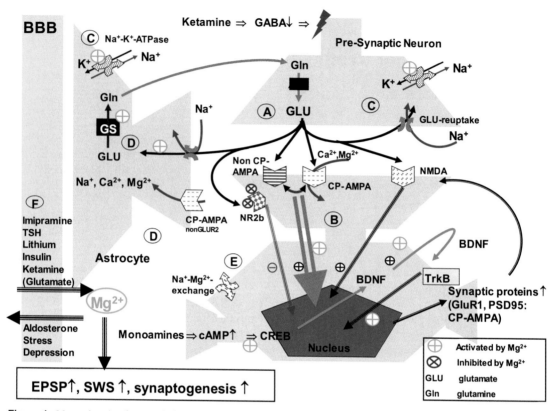

Figure 1. Magnesium involvement in ketamine-induced pathways. Ketamine administration leads to a cascade of events finally resulting in modifications of glutamatergic receptor profile and synaptogenesis. Functional consequences are increased excitatory postsynaptic potentials (EPSP) and increased slow-wave sleep; both phenomena are also induced by Mg^{2+}. In detail: (A) ketamine inhibits GABAergic interneurons and therefore activates the release of glutamate downstream in the context of partial NR2b receptor blockade. (B) AMPA mediates BDNF expression and release. BDNF activates TrkB receptor, which induces changes in gene expression. BNDF induces its own expression. NMDA receptor activation is facilitating this process.[33] This could constitute a feedforward mechanism, explaining the long-term effect of ketamine administration. Further, the expression of synaptic proteins is induced, in particular GluR1 and PSD-95, which constitute the synaptic expression of Ca^{2+}-permeable AMPA receptors (CP-AMPA).[34] Importantly, CP-AMPA are permeable for Mg^{2+}. (C) Glutamate is taken up quickly by neurons and astrocytes. This is of importance as high concentrations of glutamate can spill over to extrasynaptic NMDA receptors, which appear to be primarily from the NR2b type. Their activation can block synaptogenesis and lead to cell damage. Rapid reuptake of glutamate prevents this spillover. Glutamate reuptake is an energy-dependent process driven by the Na^+-gradient over the membrane,[35] which in itself is driven by the $Na^+–K^+$–ATPase. The $Na^+–K^+$–ATPase is dependent on Mg^{2+}, therefore increased Mg^{2+} availability supports its activity and secondarily glutamate clearance. (D) In parallel glutamate receptors at astrocytes are activated. CP-AMPA receptor activation can lead to an increase in astrocytic Ca^{2+}[35] and Mg^{2+}.[36] A potential consequence of increased Mg^{2+} in astrocytes is activation of glutamine synthetase (GS), which is Mg^{2+} dependent.[37,38] (E) A neuronal $Na^+–Mg2^+$–exchange mechanism regulates intracellular Mg^{2+} concentration. Imipramine appears to block $Na^+–Mg^{2+}$ exchange, preventing the efflux of Mg^{2+} from the neuron. (F) Magnesium uptake into the brain has been described with compounds, which are known to be efficacious in treatment-refractory depression, i.e., ketamine (via its glutamate-releasing capability), TSH, lithium, imipramine, and potentially an insulin-related mechanism (metformin, glitazones). On the other hand, stress and refractory depression are linked to lower Mg^{2+} levels in the brain, which may in part be mediated via an aldosterone mediated mechanism. Adapted with permission from Ref. 11.

action,[10] though how ketamine's activity is linked to known pathophysiological changes in depression has just begun to be understood. For example, animal models that focus on glutamatergic overactivity may be helpful in understanding vulnerability to depression and the specific biological and behavioral features that match the pharmacology of glutamatergic compounds.[11] One model in this context is the magnesium (Mg^{2+})-depletion model (Fig. 1).[12] Magnesium depletion leads to NMDA overactivity and, as a consequence, to depression and anxiety-like symptoms, neuroendocrine changes including increased cortisol levels, sleep disturbances—including a reduction of slow-wave sleep—and increased inflammatory markers. This model covers not only selected aspects but the whole variety of biological changes observed in certain patients with depression. This raises the question of the role of Mg^{2+} itself in the pathophysiology of depression; for example, can depression be induced by nutritional Mg^{2+} deficit? Moreover, Mg^{2+} may additionally be considered as a mediator of established treatments—including imipramine, lithium, and thyroid stimulating hormone (TSH)[11] treatment—that boosts improvement in refractory patients, which may open up the possibility for development of a new class of compounds based on their capability to increase neuronal and astrocyte Mg^{2+} content.

On March 25, 2013, the New York Academy of Sciences hosted the conference "Treatment-Resistant Depression" that attracted attendees from industry, academia, and governmental agencies and focused on current research that seeks to move beyond the traditional monoamine neurotransmitter–focused treatment for patients suffering from depression and related disorders.

Developing novel treatments: use of rapid-acting antidepressants and biomarkers of treatment response

Carlos A. Zarate, Jr. (National Institute of Mental Health) began the meeting with a discussion of the current horizons of development of drugs for the treatment of depression and biomarkers that could be used to evaluate patients' responses to treatment. Despite considerable effort over the last several decades, little progress has been made in developing more effective antidepressants that the current armamentarium. The first antidepressant drug, discovered by serendipity, eventually led to a multitude (over 30 specific compounds currently) of other antidepressants drugs that were little more than molecular refinements of the initial prototype drug, whose mode of action was to modulate the effects of serotonin and norepinephrine. Over the decades, drug discovery and development for depression have proceeded largely on two fronts: mimic what earlier ones do—modulate serotonin and norepinephrine—or explore new therapeutic targets for depression.

Concerning the measure of success along the first front, none of the developed compounds have demonstrated a significant advantage in terms of efficacy over earlier antidepressants, owing to the fact that all of these second and third generation drugs remain either primarily serotonergically- or noradrenergically-based (and known as selective serotonin reuptake inhibitors (SSRIs) or serotonin–norepinephrine reuptake inhibitors (SNRIs)). Rather than having greater efficacy, the drugs are, for the most part, better tolerated than the original tricyclic antidepressants and monoamine oxidase inhibitors.

Success along the second front of drug discovery, based on identifying novel therapeutic targets that result in new treatments, has unfortunately failed miserably. Several new compounds that seemed to show improvement in animal models were subsequently shown to have little effect in humans (phase I and II studies). Reasons for this lack of success in developing new and improved antidepressant have been discussed.[13,14]

One strategy that has been increasingly used in drug discovery and development strategies is the incorporation of biomarkers that either signal drug effects (target or functional engagement) or are used in treatment-response paradigms. This strategy has been encouraged by many groups, including the Institute of Medicine (IOM).[15] Technologies that have begun to be incorporated in treatment-response studies include positron emission tomography (PET); functional magnetic resonance imaging (fMRI); brain proton magnetic resonance spectroscopy (1H-MRS); neurophysiology measures such as sleep electroencephalography and magnetoencephalography (MEG); peripheral blood, plasma, and urine markers; cerebrospinal fluid (CSF); and genetics, proteomics, and metabolomics, to name several. Many of these technologies are still being refined.

One paradigm which is now being utilized to enhance drug development efforts for depression and bipolar disorder is the study of interventions that are radically distinct, in some clinically useful way, from existing treatments. The study of ketamine and scopolamine exemplify this different paradigm. First, both of these drugs have been demonstrated, in at least two controlled trials, to have much more

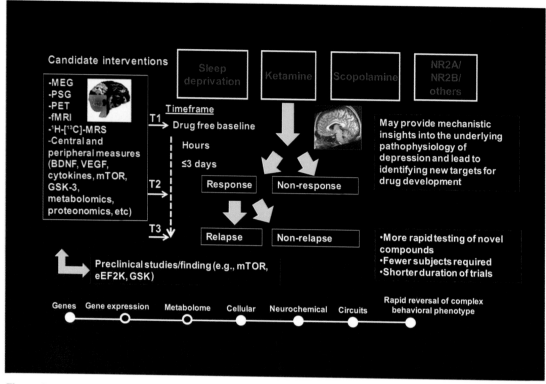

Figure 2. Conceptual framework for advancing translational findings across a systems level of biomarkers of response/relapse to develop rapid-acting antidepressants.

rapid antidepressant and antisuicidal effects than existing treatments; within a few hours for ketamine and a few days for scopolamine.[16,17] Second, ketamine is effective in patients who have failed a multitude of antidepressants as well as electroconvulsive therapy, which is the most effective current treatment. Third, these two drugs, unlike traditional antidepressants, appear to more directly modulate two distinct neurotransmitter systems, the glutamatergic and muscarinic neurotransmitter systems. Finally, because of their rapid antidepressant effects, these drugs appear to be useful tools for synchronizing possible biomarkers within a relatively short period of time, in contrast to the complexity and cost of studying existing treatments (for review see Ref. 17).

Figure 2 illustrates the new paradigm in the study of biomarkers with rapid-acting antidepressants—early work has begun to yield promising putative biomarkers predictive of rapid antidepressant response.[18,19] This model investigates a multitude of biomarkers (e.g., fMRI, MEG, PSG, brain-derived

neurotrophic factor (BDNF), SNPs) during interventions that produce rapid antidepressant clinical effects, for example, within 72 hours. Bio-signatures of response, non-response, and relapse are generated from the integration of biological findings. These results can then inform and guide drug discovery and development efforts.

Indeed, similar strategies are already being conducted on a preclinical level, where the molecular and cellular signatures of response to ketamine and scopolamine are being compared and contrasted. The antidepressant actions of scopolamine, a muscarinic antagonist, and ketamine, an NMDA antagonist, require the mammalian target of rapamycin (mTOR) signaling.[20] Such strategies that can assess the efficacy of rapid-acting antidepressants on micro and macro levels—across systems levels—may be more likely to generate important insights for developing the next generation of treatments that, hopefully, will act more rapidly and will be more effective than existing antidepressants. Such next-generation treatments may even put psychiatry on par with other areas of medicine, by allowing physicians to intervene with treatment that rapidly prevents or reverses major depressive episodes or prevents suicide by rapidly eliminating suicidal ideation.

Ketamine treatment as a new paradigm for the treatment of affective disorders: preclinical studies

Ronald S. Duman (Yale University School of Medicine) further explored the potential of ketamine as a treatment for depression and other affective disorders. Recent molecular and cellular studies have demonstrated that stress and antidepressants exert opposing effects on the expression of neurotrophic factors that result in structural alterations of neurons, including regulation of dendrite complexity and spine density in the prefrontal cortex (PFC) and the hippocampus. The deleterious effects of stress may contribute to the reduced volume of the PFC and the hippocampus in depressed patients. Conversely, the actions of antidepressants could be mediated in part by blocking or reversing the atrophy caused by stress and depression.

Recent studies have demonstrated that ketamine produces rapid (within hours) antidepressant responses in treatment-resistant depressed patients, thus addressing a major limitation of currently available agents (i.e., delayed onset of action and low response rates). The discovery that ketamine produces rapid and efficacious antidepressant effects by a mechanism completely different from conventional antidepressants (blockade of NMDA receptors (NMDARs)) represents one of the most important discoveries in the field of depression in over 50 years. Preclinical studies in rodent models have begun to unravel the molecular and cellular mechanisms underlying the rapid actions of NMDAR antagonists. Duman's presentation discussed recent work demonstrating that ketamine causes a rapid induction of spine density in the medial PFC through activation of neurotrophic factor signaling and mTOR, which regulate the translation of synaptic proteins (reviewed in Ref. 21). The roles of mTOR signaling and synaptogenesis in the response to scopolamine were also discussed by Duman.

The ability of a single dose of ketamine to cause a rapid antidepressant response suggests that this agent produces rapid effects on neuronal function, possibly through regulation of spine synapses. To directly test this possibility, Duman's group examined the influence of ketamine administration on spine number and function in layer V pyramidal neurons in slices of the PFC using single-cell patch clamp electrophysiology combined with confocal imaging of neurobiotin-labeled neurons. They found that a single dose of ketamine increased the amplitude and frequency of 5-HT– and hypocretin-induced excitatory postsynaptic currents (EPSCs) in layer V neurons.[22] Moreover, they found that ketamine increased the density of spine synapses, as well as the number of mature mushroom spines (i.e., increased spine head diameter), in both the proximal and distal dendrite branches of layer V neurons. At a behavioral level, they found that ketamine also produced a rapid antidepressant response in the forced-swim and novelty-suppressed feeding tests, and blocked anhedonia (decreased sucrose preference) caused by chronic stress exposure. The latter results provide a rigorous test and confirmation of the rapid actions of ketamine, compared to the requirement for long-term treatment (three weeks) of a typical reuptake inhibitor antidepressant to produce similar effects.

The rapid synaptogenic action of ketamine is similar to what is observed in cellular models of learning and memory, where a burst of glutamate—the major excitatory neurotransmitter in the brain—results in the release of BDNF and activation of mTOR signaling, which leads to increased translation of synaptic proteins required for synaptogenesis. Previous studies demonstrated that ketamine rapidly increases glutamate release in the PFC, possibly through inhibition of tonic firing of GABAergic interneurons, providing support for this hypothesis.[23] Duman's group found that ketamine causes a rapid (30 min) activation of mTOR signaling in the PFC, as measured by increased levels of the phosphorylated and activated forms of mTOR and S6 kinase. A role for mTOR signaling was further supported by studies demonstrating that pretreatment with rapamycin, a selective mTOR inhibitor, completely blocked the induction of spine synapses and the behavioral actions of ketamine.[22,24] Duman's group also tested the role of BDNF in mice with a knock-in of the Val66Met *BDNF* polymorphism, where the Met allele blocks activity-dependent release of BDNF. The ability of ketamine to increase spine synapses and produce a rapid antidepressant behavioral response was completely blocked in the *BDNF*[Val66Met] mice.[25] Clinical studies have demonstrated that patients carrying the $BDNF^{Val66Met}$ allele have a significantly decreased response to ketamine; thus the Val66Met polymorphism serves as a genetic marker for ketamine treatment response.

Based on these findings, Duman and colleagues have also examined the role of mTOR signaling and synaptogenesis in the actions of scopolamine, another treatment that produces rapid antidepressant actions.[19] The results thus far have demonstrated that a single dose of scopolamine also rapidly increases spine number and function in layer V neurons and increases mTOR signaling in the PFC. In addition, scopolamine produces rapid antidepressant actions in the forced swim test that are blocked by pretreatment with rapamycin. Preliminary evidence indicates that scopolamine also increases glutamate release in the PFC.

Together these findings suggest a common mechanism for the effects of rapid-acting antidepressants, including a burst of glutamate transmission that causes release of BDNF, stimulation of mTOR signaling, and increased spine number and function. The induction of spine synapses blocks or reverses the atrophy and loss of connections in cortical and limbic circuits caused by chronic stress, thereby causing reinstatement of normal circuit-connection control of mood and emotion.[21]

Targeting glutamatergic receptors

Jorge Quiroz (Roche) discussed work underway to develop antidepressant treatments targeting the glutamatergic pathway. Well-powered and adequately controlled studies have failed to demonstrate the efficacy of newer pharmacological interventions; this, in addition to high placebo-response ratios, has motivated a major withdrawal of the pharmaceutical industry from basic and clinical research in neuroscience. Despite this trend, the deepened understanding of mood disorder pathophysiology, including the better characterization of depression endophenotypes and the improvement of mechanistic and circuitry-based understanding of these diseases, has enabled investigational efforts beyond the classic monoaminergic approach for the treatment of major depression.

It is noteworthy that advances in the physiological understanding of the glutamatergic neurotransmitter system have demonstrated the modulatory controls over emotional processing and have therefore increased our capacity for neurobiological

tractability in mood disorders. Quiroz presented the rationale for targeting the mGlu2 and mGlu5 receptors, which offer novel treatment approaches that address both the depressive symptomatology and the cognitive deficits associated with depression. Consequently, Roche is currently conducting two proof-of-concept studies (clinicaltrials.gov; NCT01483469) in major depressive disorder with mGlu negative allosteric modulators as adjunctive treatment in patients with inadequate response to SSRIs and SNRIs. These therapies are being developed for the treatment of depression with the hope of improving remission rates, speed of onset, and overall quality of life for patients suffering from these devastating diseases.

Magnesium for treatment-resistant unipolar depression

Guosong Liu (Tsinghua University) shifted the discussion from antidepressant drug development toward treatments focusing on the magnesium-depletion model of depression. Patients with major depressive disorder (MDD) express strong negative emotions such as anxiety, feelings of worthlessness, helplessness, and anhedonia, as well as reduction of executive functions such as difficulty in concentrating, remembering, or making decision. Currently available drugs that target monoaminergic systems have a delayed onset of action and significant limitations in efficacy. Several studies show that MDD patients have significant synapse loss in the PFC. Since the PFC is a brain region critical for cognitive abilities and emotional control, synapse loss in the PFC might underlie the reduction of cognitive abilities and dysfunction of emotional control. Promoting synaptogenesis in the PFC might become a novel therapeutic strategy for treating MDD.

Liu's laboratory has been studying the principles governing synapse organization on the dendrite.[26] One of their findings is that elevation of extracellular Mg^{2+} concentration promotes synaptogenesis and enhances synaptic plasticity.[27] Mechanistically, they show that, *in vitro*, elevation of the extracellular concentration of Mg^{2+} can selectively reduce Ca^{2+} influx through NMDARs near resting membrane potential. This reduction of basal Ca^{2+} leads to a compensatory upregulation of NR2B-containing NMDARs, resulting in enhancement of synaptic plasticity.

To investigate the role of brain magnesium on synapse density and plasticity, Liu suggested that one needs to find a way to elevate brain Mg^{2+} effectively. Unfortunately, common magnesium compounds have poor bioavailability and, importantly, they fail to deliver magnesium to the brain in an efficient and safe manner. Taking on this challenge, Liu's group, after extensive screening in rodents, identified Mg-L-threonate (Magtein™) as a bioavailable magnesium compound that, importantly, could efficiently elevate brain magnesium concentrations.

Having demonstrated that Magtein treatment effectively increases brain magnesium levels, Liu and colleagues studied the effects of elevated brain magnesium on synapse density and plasticity in brain regions critical for cognitive abilities and emotions. They found that Magtein treatment induces a unique regional-specific pattern of action, enhancing NMDAR signaling and synaptic plasticity in the PFC and hippocampus, but not in the basolateral amygdala. The increase in synapse density and plasticity in the PFC and hippocampus correlated with enhancement of learning and memory in young and aged rats.[28,29] Interestingly, elevation of brain magnesium remain effective in preventing and reversing cognitive deficit in an Alzheimer's disease mouse model.[30] These results confirm that the level of brain magnesium is critical for cognitive functions.

Liu and colleagues have also studied the effects of elevation of brain magnesium on emotional control. Anxiety is one of major symptoms of MDD. In the clinic, cognitive therapy is effective in treating anxiety. However, many cases relapse or resist this therapy. Therefore, considerable research has been conducted to identify cognitive enhancers that promote extinction of fear memory. Fear memories are formed in the amygdala. Following extinction, fear expression is modulated by other brain regions such as the PFC and hippocampus. Therefore, enhancement of PFC and hippocampal functions without simultaneously enhancing or impairing amygdala functions would boost the efficacy of cognitive therapy for anxiety. Indeed, Magtein treatment enhances the retention of extinction and prevents spontaneous recovery and renewal of extinguished fear memories without enhancing, impairing, or erasing the amygdala-dependent fear memory.[29]

In animal models of depression-like behavior, Magtein treatment reduced the immobility time in the forced-swimming test (in a dose-dependent manner) and reversed the escape deficits in a learned-helplessness paradigm. Furthermore, Magtein treatment reduced anxiety-like behavior in three models of anxiety: elevated plus maze, novelty-suppressed feeding test, and open field.

Liu concluded by stating that if elevation of brain magnesium can enhance cognitive abilities and reduce anxiety and helplessness simultaneously, treatments that increase brain magnesium concentrations may offer a novel approach for treating depression with improved therapeutic benefits.

NMDA receptor over-reactivity as a model for unraveling biochemical pathways contributing to enhanced depression-related behavior

Simone Sartori (University of Innsbruck) further emphasized the suggestion that the glutamatergic pathways, in particular the NMDAR, may represent a novel approach to the treatment of depression, although the role of the NMDAR in the pathophysiology of affective disorders is not clear. In order to address this question, one strategy to mildly modulate central NMDAR activity is to target its voltage-dependent magnesium (Mg^{2+}) block. For example, feeding mice a low Mg^{2+}–containing diet (50–120 mg/kg food) for three weeks causes a reduction in brain and plasma Mg^{2+} levels compared with mice on a control diet;[12,30,31] a similar condition was also reported in some, though not all, patients with depression.[1] Although Mg^{2+}-deficient animals appear normal (e.g., they have shiny fur and normal gain in body weight, circadian activity rhythm, and locomotor performances[12]), their body temperature increased.[32] Mg^{2+}-deficient mice display signs of behavioral despair, as indicated by high immobility times in the forced-swim and/or tail-suspension tests, and signs of anhedonia (Sah, Sartori, Singewald, unpublished), both of which are core symptoms of patients with depression. In addition, anxiety-related behavior is increased in Mg^{2+}-deficient mice.[12,32] The pro-depressive effects of dietary-induced hypomagnesemia are observed in several mouse strains with divergent levels of depression/anxiety-related behavior including the C57/Bl6, Balb/c, and CD-1 strains, as well as in rats. While these results have been replicated in different laboratories, the anxiogenic effects of hypomagnesemia in humans seem to be more variable, reflecting the clinical setting where a comorbid anxiety disorder is diagnosed in about 60% of depressed individuals. Chronic oral administration of antidepressant drugs (paroxetine, desipramine, or St. John's Wort) normalizes the enhanced depression- and/or anxiety-related behavior displayed by Mg^{2+}-deficient mice.[12,31,32]

Despite the ubiquitous distribution of Mg^{2+} in the brain, Sartori and colleagues observed stress-induced neuronal hyperactivation in the amygdala (Singewald *et al.*, unpublished) and the paraventricular hypothalamic nucleus.[32] Using unbiased proteomics, only four out of more than 300 soluble proteins analyzed were identified as altered in the amygdala following Mg^{2+}-deficiency. Among these, changes in *N,N*–dimethylarginin-dimethylaminohydrolase 1 and manganese superoxide dismutase point toward the possibility of upregulated NMDAR/nitric oxide (NO) pathway signaling.[31] Furthermore, preliminary data from Sartori's group shows that the Mg^{2+}-deficiency–induced enhanced depression-related behavior is blocked after a single application of the NMDAR antagonist ketamine, as well as in heterozygous neuronal nitric oxide synthase knockout mice (unpublished data). These findings indicate that a hyperactive NMDAR/NO signaling pathway plays a key role in mediating the behavioral effects of diet-induced Mg^{2+} deficiency. Sartori also reported

that in Mg^{2+}-deficient mice, increased transcription of the pre–pro-corticotropin releasing hormone in the paraventricular hypothalamic nucleus leads to increased release of the stress hormones ACTH[32] and aldosterone, indicating dysfunction in the neuroendocrine stress axis that is known to control mood and emotions as well as the immune system. Correspondingly, immune responses have been shown to be altered in Mg^{2+}-deficient rodents.[1] Chronic antidepressant treatment normalizes the Mg^{2+}-deficiency–induced dysfunctions in the neuroendocrine system and NMDAR/NO signaling pathways in parallel with the observed behavioral effects.[12,31,32]

Overall, rodents with dietary-induced Mg^{2+} deficiency highly reflect patients with depression in terms of deranged physiological homeostasis, behavior, neurobiology, and antidepressant treatment responses. Despite or even because of the multifunctional properties of Mg^{2+} in the brain,[1] the Mg^{2+} deficiency is a model for studying neurobiological mechanisms beyond the monoaminergic systems leading from chronic (over-) activation of the NMDAR and stress pathways to enhanced depression- and anxiety-related behavior. This is pivotal knowledge for the development of novel pharmacotherapies. However, because no novel antidepressant with greatly improved pharmacological properties has been approved by the FDA in the last 30 years, refinement of drug development at early preclinical stages is recommended by using

Table 1. Comparison of ketamine- and magnesium-induced changes with reference to those observed in major depression

	Major depression (human)	Magnesium administration	Ketamine
PFC-NR1	⇔[39]	⇔[26]	⇑[40]
PFC-NR2A	⇓[39]	⇔[26]	⇔[40]
PFC-NR2B	⇓[39]	⇑[26]	⇑[40,41]
Cortical (PFC)–GluR1	?	?	⇑[24,42] (after CUS)
Cortical (PFC)–GluR2	⇑[43]	?	?
PFC–BNDF	?	⇑[26]	⇑[44]
PFC–PSD-95	⇓[39]	?	⇑[24] (after CUS)
Cortical	⇓[45,46]	⇑[47]	⇑[44,48]
P-CREB		⇔[26]	
PFC/ACC–glutamine/GS activity	⇓[46]	⇑[49]	⇑[50]
Evoked synaptic potentials (cortex)	?	⇑[26]	⇑[24] (after CUS)
Cortical synaptogenesis	?	⇑[26]	⇑[24] (after CUS)

Adapted with permission from Ref. 11. CUS, chronic unpredictable stress.

well-characterized model organisms with high face, construct, and predictive validity to the human disorder. Accordingly, the rodent Mg^{2+}-deficiency model is clinically relevant for testing potential novel drug targets with fast-acting antidepressant properties. For an overview of the discussed effects, see Table 1.

Conclusion

While the standard of care for people suffering from clinical depression has not changed significantly since the introduction of SSRIs and SNRIs decades ago, current research has explored novel paradigms with the potential to develop new treatments to address the unmet need in depression patients. This research has been facilitated by the development of biomarkers and technologies that signal drug effects or can be used to gauge patient responses. The rapid response observed in patients taking the muscarinic antagonist scopolamine and the glutamatergic antagonist ketamine revealed these pathways to be potent targets for antidepressant drug development. Glutamatergic antagonism reverses the atrophy and loss of connections in cortical and limbic circuits caused by chronic stress, reestablishing normal circuit-connection control of mood and emotion. While ongoing developmental research aims to develop optimal antagonists and allosteric inhibitors of the glutamatergic pathway, other work is exploring magnesium homeostasis in the brain: magnesium levels modulate glutamatergic activity, and low brain magnesium levels correspond to animal models of depression, while high levels increase synaptogenesis and synaptic plasticity. The synthesis of research exploring how modifying these systems changes the structure and function of neurons and pathways with development of treatments targeting these pathways offers hope for improving the speed of response, remission, and quality of life for patients suffering from depression.

Conflicts of interest

The authors declare no conflicts of interest. J.Q. is employed by Roche, and H.M. is employed by Covance.

References

1. Eby, G.A., K.L. Eby & H. Murck. 2011. Magnesium and major depression. In *Magnesium in the Central Nervous System*, R. Vink & M. Nechifor, Eds.: 313–330. Adelaide: University of Adelaide Press.

2. Vale, S., M.A. Espejel & J.C. Dominguez. 1971. Amantadine in depression. *Lancet* **2:** 437.

3. Paul, I.A. & P. Skolnick. 2003. Glutamate and depression: clinical and preclinical studies. *Ann. N. Y. Acad. Sci.* **1003:** 250–272.

4. Watkins, J.C. & D.E. Jane. 2006. The glutamate story. *Br. J. Pharmacol.* **147:** S100–S108.

5. Schildkraut, J.J. 1965. The catecholamine hypothesis of affective disorders: a review of supporting evidence. *Am. J. Psychiatry* **122:** 509–522.

6. Kirsch, I., T.J. Moore, A. Scoboria & S.S. Nicholls. 2002. The emperors´s new drugs: an analysis of antidepressant medication data submitted to the U.S. food and drug administration. *Prevention and Treatment* **5:** 1–10.

7. Trivedi, M.H., A.J. Rush, S.R. Wisniewski, *et al.* 2006. Evaluation of outcomes with citalopram for depression using measurement-based care in STAR*D: implications for clinical practice. *Am. J. Psychiatry* **163:** 28–40.

8. Berman, R.M., A. Cappiello, A. Anand, *et al.* 2000. Antidepressant effects of ketamine in depressed patients. *Biol. Psychiatry* **47:** 351–354.

9. Zarate, C.A., Jr., J.B. Singh, P.J. Carlson, *et al.* 2006. A randomized trial of an *N*-methyl-D-aspartate antagonist in treatment-resistant major depression. *Arch. Gen. Psychiatry* **63:** 856–864.

10. Duman, R.S., N. Li, R.J. Liu, *et al.* 2012. Signaling pathways underlying the rapid antidepressant actions of ketamine. *Neuropharmacology* **62:** 35–41.

11. Murck, H. 2013. Ketamine, magnesium and major depression – From pharmacology to pathophysiology and back. *J. Psychiatr. Res.* **47:** 955–965.

12. Singewald, N., C. Sinner, A. Hetzenauer, *et al.* 2004. Magnesium-deficient diet alters depression- and anxiety-related behavior in mice–influence of desipramine and Hypericum perforatum extract. *Neuropharmacology* **47:** 1189–1197.

13. Miller, G. 2010. Is pharma running out of brainy ideas? *Science* **329:** 502–504.

14. Hyman, S. 2012. Revolution stalled. *Sci. Transl. Med.* **4:** 155

15. Institute of Medicine. 2008. Neuroscience Biomarkers and Biosignatures: Converging Technologies, Emerging Partnerships, Workshop Summary. Washington (DC).

16. Zarate, C.A., Jr., N.E. Brutsche, *et al.* 2012. Replication of ketamine's antidepressant efficacy in bipolar depression: a randomized controlled add-on trial. *Biol. Psychiatry* **71:** 939–946.

17. Zarate, C.A., Jr., D.C. Mathews & M.L. Furey. 2013. Human biomarkers of rapid antidepressant effects. *Biol. Psychiatry* **73:** 1142–1155.

18. Salvadore, G., B.R. Cornwell, *et al.* 2010. Anterior cingulate desynchronization and functional connectivity with the amygdala during a working memory task predict rapid antidepressant response to ketamine. *Neuropsychopharmacology* **35:** 1415–1422.

19. Furey, M.L., W.C. Drevets, *et al.* 2013. Potential of pretreatment neural activity in the visual cortex during emotional processing to predict treatment response to scopolamine in major depressive disorder. *JAMA Psychiatry* **70:** 280–290.

20. Voleti, B., A. Navarria, *et al.* in press. Scopolamine rapidly increases mammalian target of rapamycin complex 1 signaling, synaptogenesis, and antidepressant behavior response. *Biol. Psychiatry.*

21. Duman, R.S. & G.K. Aghajanian. 2012. Synaptic dysfunction in depression: Novel therapeutic targets. *Science* **338:** 68–72.

22. Li, N., B. Lee, R.J. Liu, *et al.* 2010. mTOR-dependent synapse formation underlies the rapid antidepressant effects of NMDA antagonists. *Science* **329:** 959–964.

23. Homayoun, H. & B. Moghaddam. 2007. NMDA receptor hypofunction produces opposite effects on prefrontal cortex interneurons and pyramidal neurons. *J. Neurosci.* **27:** 11496–11500.

24. Li, N., R.J. Liu, J.M. Dwyer, *et al.* 2011. Glutamate N-methyl-D-aspartate receptor antagonists rapidly reverse behavioral and synaptic deficits caused by chronic stress exposure. *Biol. Psychiatry* **69:** 754–761.

25. Liu, R.J., F.S. Lee, X.-Y. Li, *et al.* 2012. BDNF Met allele decreases the density and function of spine/synapses in the prefrontal cortex and blocks the effects of ketamine. *Biological Psychiatry* **71:** 996–1005.

26. Abumaria, N., B. Yin, L. Zhang, *et al.* 2011. Effects of elevation of brain magnesium on fear conditioning, fear extinction, and synaptic plasticity in the infralimbic prefrontal cortex and lateral amygdala. *J. Neurosci.* **31:** 14871–14881.

27. Li, W., J. Yu, Y. Liu, *et al.* 2013. Elevation of brain magnesium prevents and reverses cognitive deficits and synaptic loss in Alzheimer's disease mouse model. *J. Neurosci.* **33:** 8423–8441.

28. Liu, G. 2004. Local structural balance and functional interaction of excitatory and inhibitory synapses in hippocampal dendrites. *Nat. Neurosci.* **7:** 373–379.

29. Slutsky, I., S. Sadeghpour, B. Li & G. Liu. 2004. Enhancement of synaptic plasticity through chronically reduced Ca(2+) flux during uncorrelated activity. *Neuron* **44:** 835–849.

30. Whittle, N. *et al.* 2011. Changes in brain protein expression are linked to magnesium restriction-induced depression-like behavior. *Amino Acids* **40:** 1231–1248.

31. Eby, G.A., K.L. Eby & H. Murck. 2011. Magnesium and major depression. In *Magnesium in the Central Nervous System*, R. Vink & M. Nechifor, Eds.: 303–312. Adelaide: University of Adelaide Press.

32. Sartori, S.B. *et al.* 2012. Magnesium deficiency induces anxiety and HPA axis dysregulation: modulation by therapeutic drug treatment. *Neuropharmacology* **62:** 304–312.

33. Xiong, H., T. Futamura, H. Jourdi, *et al.* 2002. Neurotrophins induce BDNF expression through the glutamate receptor pathway in neocortical neurons. *Neuropharmacology* **42:** 903–912.

34. Fortin, D.A., T. Srivastava, D. Dwarakanath, *et al.* 2012. Brain-derived neurotrophic factor activation of CaM-kinase kinase via transient receptor potential canonical channels induces the translation and synaptic incorporation of GluA1-containing calcium-permeable AMPA receptors. *J. Neurosci.* **32:** 8127–8137.

35. Magistretti, P.J. 2009. Role of glutamate in neuron-glia metabolic coupling. *Am. J. Clin. Nutr.* **90:** 875S–880S.

36. Muller, A., D. Gunzel & W.R. Schlue. 2003. Activation of AMPA/kainate receptors but not acetylcholine receptors causes Mg^{2+} influx into Retzius neurones of the leech Hirudo medicinalis. *J. Gen. Physiol.* **122:** 727–739.

37. Greenberg, J. & N. Lichtenstein. 1959. Effect of manganous and magnesium ions concentration on glutamine synthetase and glutamotransferase of sheep brain. *J. Biol. Chem.* **234:** 2337–2339.

38. Maurizi, M.R., H.B. Pinkofsky, P.J. McFarland & A. Ginsburg. 1986. Mg^{2+} is bound to glutamine synthetase extracted from bovine or ovine brain in the presence of L-methionine-S-sulfoximine phosphate. *Arch. Biochem. Biophys.* **246:** 494–500.

39. Feyissa, A.M., A. Chandran, C.A. Stockmeier & B. Karolewicz. 2009. Reduced levels of NR2A and NR2B subunits of NMDA receptor and PSD-95 in the prefrontal cortex in major depression. *Prog. Neuropsychopharmacol. Biol. Psychiatry* **33:** 70–75.

40. Chatterjee, M., R. Verma, S. Ganguly & G. Palit. 2012. Neurochemical and molecular characterization of ketamine-induced experimental psychosis model in mice. *Neuropharmacology*

41. Burgdorf, J., X.L. Zhang, K.L. Nicholson, *et al.* 2013. GLYX-13, a NMDA receptor glycine-site functional partial agonist, induces antidepressant-like effects without ketamine-like side effects. *Neuropsychopharmacology* **38:** 729–742.

42. Yamada, S., M. Yamamoto, H. Ozawa, *et al.* 2003. Reduced phosphorylation of cyclic AMP-responsive element binding protein in the postmortem orbitofrontal cortex of patients with major depressive disorder. *J. Neural Transm.* **110:** 671–680.

43. Teyssier, J.R., S. Ragot, Chauvet-J.C. Gelinier, *et al.* 2011. Activation of a DeltaFOSB dependent gene expression pattern in the dorsolateral prefrontal cortex of patients with major depressive disorder. *J. Affect Disord.* **133:** 174–178.

44. Reus, G.Z., R.B. Stringari, K.F. Ribeiro, *et al.* 2011. Ketamine plus imipramine treatment induces antidepressant-like behavior and increases CREB and BDNF protein levels and PKA and PKC phosphorylation in rat brain. *Behav. Brain Res.* **221:** 166–171.

45. Dwivedi, Y., J.S. Rao, H.S. Rizavi, *et al.* 2003. Abnormal expression and functional characteristics of cyclic adenosine monophosphate response element binding protein in postmortem brain of suicide subjects. *Arch. Gen. Psychiatry* **60:** 273–282.

46. Choudary, P.V., M. Molnar, S.J. Evans, *et al.* 2005. Altered cortical glutamatergic and GABAergic signal transmission with glial involvement in depression. *Proc. Natl. Acad. Sci. USA* **102:** 15653–15658.

47. Huang, C.Y., Y.F. Liou, S.Y. Chung, *et al.* 2010. Role of ERK signaling in the neuroprotective efficacy of magnesium sulfate treatment during focal cerebral ischemia in the gerbil cortex. *Chin. J. Physiol.* **53:** 299–309.

48. Shu, I., T. Li, S. Han, *et al.* 2012. Inhibition of neuron-specific CREB dephosphorylation is involved in propofol and ketamine-induced neuroprotection against cerebral ischemic injuries of mice. *Neurochem. Res.* **37:** 49–58.

49. Maurizi, M.R., H.B. Pinkofsky, P.J. McFarland & A. Ginsburg. 1986. Mg^{2+} is bound to glutamine synthetase extracted from bovine or ovine brain in the presence of L-methionine-S-sulfoximine phosphate. *Arch. Biochem. Biophys.* **246:** 494–500.

50. Rowland, L.M., J.R. Bustillo, P.G. Mullins, *et al.* 2005. Effects of ketamine on anterior cingulate glutamate metabolism in healthy humans: a 4-T proton MRS study. *Am. J. Psychiatry* **162:** 394–396.

Ann. N.Y. Acad. Sci. ISSN 0077-8923

ANNALS OF THE NEW YORK ACADEMY OF SCIENCES
Issue: *Glycobiology of the Immune Response*

Carbohydrate recognition in the immune system: contributions of neoglycolipid-based microarrays to carbohydrate ligand discovery

Ten Feizi

The Glycosciences Laboratory, Department of Medicine, Imperial College London, London, United Kingdom

Address for correspondence: Ten Feizi, The Glycosciences Laboratory, Department of Medicine, Imperial College London, Hammersmith Campus, Burlington Danes Building, Du Cane Road, W12 0NN, UK. t.feizi@imperial.ac.uk

Oligosaccharide sequences in glycomes of eukaryotes and prokaryotes are enormously diverse. The reasons are not fully understood, but there is an increasing number of examples of the involvement of specific oligosaccharide sequences as ligands in protein–carbohydrate interactions in health and, directly or indirectly, in every major disease, be it infectious or noninfectious. The pinpointing and characterizing of oligosaccharide ligands within glycomes has been one of the most challenging aspects of molecular cell biology, as oligosaccharides cannot be cloned and are generally available in limited amounts. This overview recounts the background to the development of a microarray system that is poised for surveying proteomes for carbohydrate-binding activities and glycomes for assigning the oligosaccharide ligands. Examples are selected by way of illustrating the potential of "designer" microarrays for ligand discovery at the interface of infection, immunity, and glycobiology. Particularly highlighted are sulfo-oligosaccharide and gluco-oligosaccharide recognition systems elucidated using microarrays.

Keywords: oligosaccharides; neoglycolipids; carbohydrate microarrays; carbohydrate ligands

The diversity of oligosaccharide sequences in glycomes and the challenges for ligand discovery

Oligosaccharide sequences of glycoproteins, glycolipids, proteoglycans, and polysaccharides are enormously diverse. In mammalian glycomes, for example, they range from 2 to more than 200 monosaccharides in length joined by differing linkages, and some are modified, for example, by sulfation. Some oligosaccharide sequences are widely distributed in different cell types, whereas others are restricted to certain cell types; others are genetically determined (by glycosyltransferase genes). Examples of the latter are the blood group antigens and species-specific oligosaccharide antigens; compatibilities of these are major considerations in blood transfusion and xenotransplantation. Oligosaccharide sequences have a regulated expression at different stages of embryonic development and cellular differentiation, and they are often aberrantly expressed in cancer cells, such that they behave as differentiation antigens and cancer-associated antigens.[1]

The reasons for the diversity of oligosaccharides sequences within glycomes are unknown. There is an elegant hypothesis that the diversity may, in part at least, represent an evolutionary adaptation and diversification, under natural selection pressures in the course of endogenous and host–pathogen interactions.[2,3] Indeed, there is increasing evidence that specific oligosaccharide sequences are directly involved as ligands in protein–carbohydrate interactions, *in vivo*, in health and, directly or indirectly, in the majority of diseases be they infectious or noninfectious. Examples of oligosaccharides as recognition elements in health include the folding of newly synthesized proteins in the endoplasmic reticulum, the routing of proteins (lysosomal enzymes) to their correct destinations in cells, the clearance of aged (asialo)-glycoproteins from serum, and the endocytosis of fungal pathogens for destruction. As discussed by authors in a recent focus issue on glycobiology of the immune response,

doi: 10.1111/nyas.12210

(*Ann. N. Y. Acad. Sci.* **1253:** 1–221) protein–carbohydrate interactions have crucial roles in the regulation of inflammatory processes and in mechanisms of immunity (innate and acquired). Notable examples of oligosaccharides as ligands in disease processes are the attachment of adhesive proteins of pathogens to selected oligosaccharides of host cells at initial stages of infection.[4] This is doubtless the tip of the iceberg.

With screening technologies such as carbohydrate microarrays, entirely novel and unsuspected carbohydrate recognition systems are being discovered, as for example, recognition of di-glucosyl-high-manose *N*-glycans by the endoplasmic recognition protein malectin.[5] Another observation from microarray screening analyses for Simian Virus 40 (SV40) recognition was the finding that that the *N*-glycolyl analogue of ganglioside GM1 is preferentially bound.[6] Unlike the *N*-acetyl analogue of GM1 ganglioside, which is found in humans and many mammals, the *N*-glycolyl GM1 is not synthesized by humans and is characteristic of simian species and other nonhuman mammals. The paucity of this ganglioside in humans may indeed be an example of evolutionary genetics of the type discussed above whereby humans have evolved not to synthesize *N*-glycolyl analogue of *N*-acetylneuraminic acid, and thereby develop at least a partial barrier to infection.

Identifying and determining the sequences of oligosaccharides that are ligands in biological systems that operate through carbohydrate recognition has been a challenging aspect of cell and molecular biology; this is because oligosaccharides cannot be cloned, they can generally be accessed in very limited amounts; and the affinities of interactions with recognition proteins are in most cases low. This review is focused on a technology, the neoglycolipid (NGL) technology, first introduced with my colleagues in 1985 in order to be able to carry out microscale direct-binding studies using oligosaccharides derived from glycoproteins.[7,8] Here, I describe briefly the technology, the contributions before and following its miniaturization, and conversion to a state-of-the-art carbohydrate microarray system that is poised for ambitious glycomic analyses of carbohydrate recognition, and I dwell on some observations at the interface of immunity and glycobiology.

NGL technology

The impetus for developing the NGL technology—the principles

In classical studies that led to assignments of the structures of the major blood group antigens A, B, and H, the amounts of purified oligosaccharides needed were enormous. These were hemagglutination-inhibition or inhibition-of-precipitation assays requiring milligram mounts of purified oligosaccharide per well or per tube.[9,10] The tricks of the trade at that time were to identify abundant sources of the antigens, such as the blood group antigen–rich mucin glycoproteins from ovarian cystadenomas. Similarly, these mucins served as sources of oligosaccharides, to characterize carbohydrate differentiation antigens such as the murine stage specific embryonic antigen, SSEA-1[11] and the human myeloid cell-specific antigen,[12] now termed CD15. The large amounts of oligosaccharides required in these types of analysis meant that ligands for very few other recognition systems involving the sugar chains of glycoproteins could be readily tackled.

Clearly, there was a need for a microtechnique, and this was the impetus for developing the NGL technology, which involves microscale conjugation of oligosaccharides via their reducing ends to a lipid molecule (Fig. S1). We selected an aminophospholipid and reductive amination as the conjugation principle. The resulting NGLs, with their amphipathic properties could be immobilized on matrices in a highly desirable, clustered state for direct-binding analyses. Moreover, the NGLs derived from oligosaccharide mixtures lent themselves well to binding experiments after resolution on thin layer chromatograms, similarly to glycolipids.[13,14] The NGLs were found to have excellent ionization properties in mass spectrometry, thus the technology was elaborated to include chromatogram-binding experiments in conjunction with mass spectrometry[15] (Fig. S2). The reductively generated NGLs serve well with oligosaccharides on which the recognition motifs are at the periphery of the chain. However, the core monosaccharide is largely sacrificed due to ring opening during conjugation to lipid. An alternative conjugation by the amino-oxy principle results in NGLs with a significant proportion of ring-closed cores[16] (Fig. S1); this is advantageous

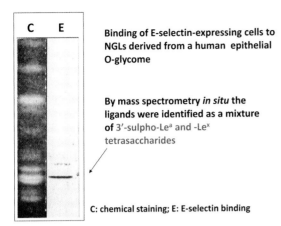

Binding of E-selectin-expressing cells to NGLs derived from a human epithelial O-glycome

By mass spectrometry *in situ* the ligands were identified as a mixture of 3′-sulpho-Le^a and -Le^x tetrasaccharides

C: chemical staining; E: E-selectin binding

Figure 1. Oligosaccharide ligand discovery for the endothelium-leukocyte adhesion molecule E-selectin in a human epithelial O-glycome (adapted from Ref. 17).

for short oligosaccharides. Moreover, this also caters for recognition systems where, for example, the *N*-glycan core monosaccharide is involved as a part of the recognition motif.[16] Although the NGL technology was originally established for oligosaccharides derived from glycoproteins, it is equally applicable to those derived from diverse glycoconjugates or synthesized chemically.[15]

Contributions of NGL technology (1985–2002) before miniaturization for microarrays

An example of the power of the NGL technology for pinpointing ligands within glycomes is the discovery of sulphated Lewis^a and sulphated Lewis^x as ligands for E-selectin within the highly heterogeneous O-glycome of an epithelial mucin[17] (Fig. 1). Other contributions (Fig. S3) included elucidations of differentiation antigens: the neural induction antigen L5 (Ref. 18) and the scrapie lesion antigen 10E4 (Ref. 19), and application of the technology in bacterial adhesion studies.[20] The hitherto unknown O-mannosyl glycans were shown to be the sole O-glycan carriers of the HNK-1 antigen in the brain[21] and to constitute about 30% of O-glycans in the brain.[22] We have reviewed elsewhere the assignments of ligands for endogenous carbohydrate-binding proteins that are receptors of the immune system, enabled by the NGL technology[17,23–39] they are also given in (Fig. S3). A remarkable cooperativity with effective display of the two ligands of P-selectin: sialyl-Le^x and sulfotyrosine in lipid-linked form on liposomes was also demonstrated.[40]

A highlight was the assignment of high-mannose *N*-glycans at Asn-917 of the complement glycoprotein, C3, as ligands for conglutinin (the first known C-type mammalian lectin).[26] Conglutinin does not bind to the native C3 molecule. Only after two small peptides are cleaved off from C3 in the complement cascade, can conglutinin bind to the resulting glycopeptide iC3b (Fig. S4). A marked conformational change is known to occur after the proteolytic cleavages. These findings highlighted the profound influence of the polypeptide on glycan presentation for recognition; furthermore they demonstrated that commonly occurring carbohydrate chains can mediate biological specificity in particular body compartments.

A carbohydrate microarray system based on the NGL technology—poised for deciphering the *meta*-glycome

A state-of-the-art carbohydrate microarray system based on the NGL technology

Among special features of the NGL technology are that it is applicable to minute amounts of starting oligosaccharides,[41] and analyses with glycosylceramides can be performed in parallel with NGLs.[8] The technology lends itself well to generating designer microarrays from targeted sources, for example, from polysaccharides following their partial depolymerization,[42] and from *N*- and *O*-glycomes,[43] and in principle from the *meta*-glycome (a term I use, with instigation from Bernard Henrissat, to mean glycomes of diverse origins).

In 2002, we introduced an oligosaccharide microarray system based on the NGL technology.[44] Following development, this is now a state-of-the-art system (Fig. 2).[42,45,46] Our library of sequence-defined probes currently numbers ~800, and encompasses a variety of mammalian type sequences, representative of *N*-glycans (high-mannose-type and neutral and sialylated complex-type), peripheral regions of O-glycans; blood group antigen-related sequences (A, B, H, Lewis^a, Lewis^b, Lewis^x, and Lewis^y) on linear or branched backbones and their sialylated and/or sulfated analogs; linear and branched poly-*N*-acetyllactosamine sequences; gangliosides, oligosaccharide fragments of glycosaminoglycans and polysialic acid. The arrays also include designer oligosaccharide probes of microbial and plant-derived homo-oligomers of glucose and of other monosaccharides (Fig. S5).

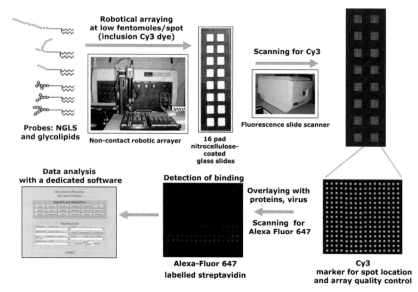

Figure 2. The neoglycolipid (NGL)-based microarray platform.

Essential to the analysis system has been the development of a database for storage and scrutiny of microarray data.[47] This holds all of the microarray data, the experimental conditions, and information on saccharide probes and proteins. There is an associated interactive software for presentation of microarray data, displaying data as tables, charts, or "matrices," selective data retrieval, filtering, sorting, and deep mining of every data point.

Contributions of NGL microarrays at the interface of immunity and glycobiology

Carbohydrate microarrays are revolutionizing the molecular dissection of protein–carbohydrate interaction.[45,48] Salient ligand assignments made with the NGL technology since proof-of-concept microarray experiments in 2002[44] are given in Figure S6. They include carbohydrate ligands for endogenous proteins,[5,42,49–53] and the adhesive proteins of infective agents,[54–57] as well as for neutralizing antibodies: antifungal[58,59] and anti-HIV.[60,61] I have selected below, by way of illustration, three topics at the interface of immunity and glycobiology.

The gluco-oligosaccharide recognition by C-type lectin-like receptor Dectin-1

Dectin-1, the major receptor of the innate immune system against fungal pathogens is a C-type lectin-like protein on leukocytes.[62] It has been reported

to interact with a subset of T lymphocytes, but the main proven function for Dectin-1 is the mediation of phagocytosis and inflammatory mediator release in innate immunity to fungal pathogens. Although lacking in residues involved in calcium ligation that mediate carbohydrate binding by classical C-type lectins, Dectin-1 binds zymosan, a particulate β-glucan–rich extract of *Saccharomyces cerevisiae*; and binding is inhibited by polysaccharides rich in β1–3- or both β1–3- and β1–6-linked glucose.[63] In order to prove that Dectin-1 binds to carbohydrates rather than to any polypeptides associated with the fungal cell walls and to identify the ligand structure(s), we generated designer probes, that is, NGLs, derived from oligosaccharides up to 13 mers isolated from partially depolymerized ligand-positive and ligand-negative glucan polysaccharides. These glucan probes were incorporated into microarrays.[42] Upon probing the microarrays, the ligands for Dectin-1 were found to be unusually long, 10-mer or longer, β1–3-linked gluco-oligosaccharides. The β1–6-linked analogs were not bound (Fig. 3).[42]

The long chain requirement for Dectin-1 binding is apparent also with on-array inhibition experiments in which we evaluated activities of β1–3-linked 7, 11, and 13 mers in solution as inhibitors of Dectin-1 binding. Only the 11 and 13 mers were inhibitory (unpublished). Thus, the long chain requirement is not related to the mode of

Ann. N.Y. Acad. Sci. 1292 (2013) 33–44 © 2013 The Authors. *Annals of the New York Academy of Sciences* published by Wiley Periodicals Inc. on behalf of The New York Academy of Sciences.

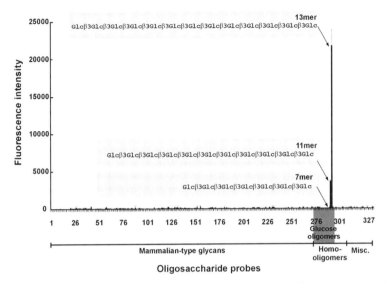

Figure 3. Microarray analysis of Dectin-1 showing binding only to β1,3-linked glucose oligomers >10 mers (from Ref. 45).

oligosaccharide presentation in the microarray, and points to the requirement for a conformational epitope, formed by the long β1–3-linked chains, analogous to the conformational epitopes on polysialic acid chains.[64] This contrasts with data with other glucan recognizing proteins, such as antibodies (discussed later).[58,59] These can bind short chains in the series. We followed up the biological significance, and observed that the ligand-positive oligosaccharides analyzed as NGLs on liposomes share the bioactivities of the intact polysaccharides.[42] Thus far, microarray analyses using mammalian type glycans have not revealed an endogenous ligand for Dectin-1. Thus, the nature of an endogenous ligand believed to be expressed on T lymphocytes is an open question and will remain under review as new oligosaccharide probes become available.

Knowing that glucan polysaccharides can act as immunomodulators, we are broadening the repertoire of designer probes from glucomes to encompass all naturally occurring glucose linkages to study recognition by other carbohydrate-binding proteins of the immune system.

The gluco-oligosaccharide determinants of vaccine-induced antifungal antibodies with therapeutic potential

The humoral response to glucans is another area of immuno-glycobiological interest that has potential for the development of therapeutic antibodies to improve the control of fungal infections in immunodeficient people. Anti-β-glucan antibodies elicited in mice by a laminarin-conjugate vaccine confer protection to mice challenged with major fungal pathogens such as *Candida albicans, Aspergillus fumigatus*, and *Cryptococcus neoformans.*[58] A monoclonal antibody (mAb) 2G8 of IgG2b subclass has been selected for further study as it confers substantial protection against mucosal and systemic candidiasis in passive transfer experiments in rodents. The binding of this antibody to the β1,3-linked gluco-oligosaccharides is shown in Figure S7. Note that this antibody contrasts with Dectin-1, in showing binding signals with short oligosaccharide probes, as well as longer sequences.

Rather than opsonization of fungal cells, the mechanism of action of mAb 2G8 has been correlated with the antibody's targeting of the heterogeneous, polydisperse, high molecular weight cell wall, and secretory components of *C. albicans.* Two of these were identified as the GPI-anchored cell wall proteins Als3 and Hyr1. In addition, mAb 2G8 inhibited *in vitro* two critical virulence attributes of the fungus: hyphal growth and adherence to human epithelial cells.[58] Data also suggest that the antivirulence properties of the mAb 2G8 antibody may be linked to its capacity to recognize β-glucan epitope(s) on some cell wall components that exert critical functions in fungal cell wall structure and adherence to host cells.[58]

Transfer of this antifungal mAb into the clinical setting would potentially enable the control of the most frequent fungal infections in immunocompromised patients. With this aim, two chimeric mouse-human derivatives of mAb 2G8, in the form of complete IgG or scFv-Fc, have been generated, transiently expressed in *Nicotiana benthamiana* plants and purified from leaves in high yields (approximately 50 mg/kg of plant tissues).[59] Both recombinant antibodies fully retain the antifungal activities of the parent murine mAb against *C. albicans*, and like the parent antibody, they bind preferentially to the β1,3-linked glucan probes. Both the IgG and the scFv-Fc promote killing of *C. albicans* by isolated, human polymorphonuclear neutrophils

in *ex vivo* assays, and confer significant antifungal protection in animal models of systemic or vulvovaginal *C. albicans* infection. Using new approaches to humanizing their *N*-glycans, these plant antibodies hold promise as therapeutic tools to confer protection against a broad range of fungal pathogens.

Sulfo-oligosaccharide ligands: modulatory effects of sulfation on recognition by selectins, siglecs, langerin, and influenza viruses

As large numbers of oligosaccharides are probed in parallel and without bias, microarrays of oligosaccharides in glycomes, and of an increasing number

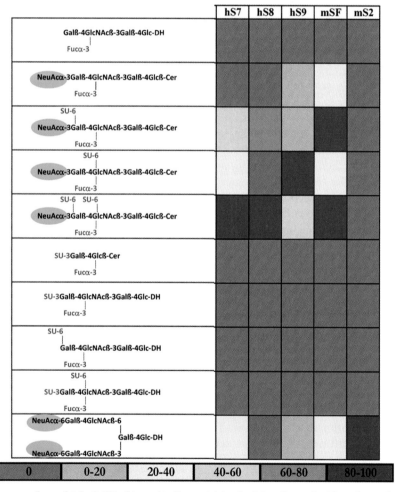

Figure 4. Microarray analyses of siglec IgG Fc chimera binding to sialyl and sulpho-oligosaccharide probes. Colors depict relative binding strengths calculated as a percentage of maximum binding to the best ligand for each siglec: three human (h) and two murine (m) (from Ref. 51).

Ann. N.Y. Acad. Sci. 1292 (2013) 33–44 © 2013 The Authors. *Annals of the New York Academy of Sciences* published by Wiley Periodicals Inc. on behalf of The New York Academy of Sciences.

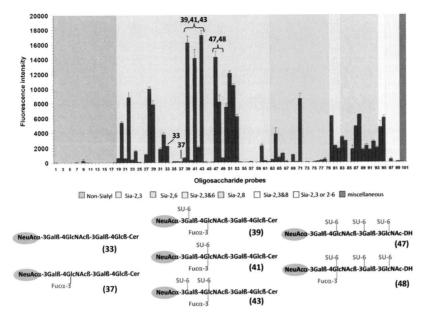

Figure 5. Microarray analysis of murine Siglec-E (from Ref. 53).

of sequence-defined oligosaccharides, have a way of revealing hitherto unsuspected ligands. A recent unpredicted "hit" with mab AE3, which binds to diverse epithelial cancer tissues and to high molecular weight mucins that they shed, is the sulfoglycolipid SM1a.[46] The carbohydrate sequence of SM1a has not been described on mucin glycoproteins. (We are pursuing the expression of the AE3 antigen on O-glycans in our Alliance of Glycobiologists NCI grant: http://glycomics.cancer.gov/teams/fukuda.)

An early example of an unpredicted finding within an epithelial O-glycome was the binding of the endothelial adhesion molecule, E-selectin to sulfo-oligosaccharide sequences of Lewisª/Lewisˣ type (with sulfate rather than sialic acid) at the terminal galactose (Fig. 1).[17] This showed that the combining site of E-selectin could accommodate equally well sulfate and the carboxyl group of sialic acid.[65] The same was later shown for L- and P-selectins.[33,34]

In the case of siglec recognition, however, microarray analyses have shown that the interchangeable role for sulfate and the sialic acid carboxyl group does not apply. As their name implies, siglecs absolutely require terminal sialic acid for binding (Fig. 4). Sulfate on sialyl oligosaccharide sequences can, nevertheless, modulate the strength of siglec binding.[51] For example, the presence of sulfate 6-linked to the adjoining galactose or 6-linked

to the inner monosaccharide, N-acetylglucosamine, can variously enhance or suppress interactions of the sialyl ligands of siglecs (Figs. 4 and 5). A detailed review of the modulatory effects of sulfation on oligosaccharide ligand recognition is out of the scope of this article. Suffice it to highlight that with well characterized, chemically synthesized glycolipids it was shown that with L-selectin, 6-linked sulfate at subterminal N-acetyl glucosamine (popularly termed 6-sulfo-sialyl-sialyl-Leˣ), enhances the binding to the sialyl-Leˣ sequence and creates the preferred ligand for this selectin.[66] This was corroborated by subsequent studies.[67] In contrast, 6-linked sulfate at the galactose, 6′-sulfo-sialyl-sialyl-Leˣ, suppresses L-selectin binding while creating the preferred ligand for langerin.[49,66]

With the NGL-based microarray system, we have also observed an influence of sulfation of sialyl oligosaccharides on receptor-binding by the pandemic influenza A(H1N1) 2009 (pdm) virus.[56,57] Although, as with seasonal influenza viruses, the strongest binding, overall, of the pdm viruses was to α2,6-linked sialyl sequences that are abundantly expressed in the upper respiratory tract, we detected binding of pdm viruses also to α2,3-linked sialyl sequences with sulfate (or fucose) at penultimate N-acetylglucosamine of the type that occur lower down in the respiratory tract (Fig. 6). The intensity

Probe designation	Oligosaccharide sequence	'wild type' 222D		222G mutants		
		Mol/09	Ham/09	Ham/09 /'egg'	Lvi/09	Nor/09
LSTa	NeuAcα-3Galß-3GlcNAcß-3Galß-4Glc-DH	-*	-	6,988	10,603	8,229
Sialyl keratan sulphate	NeuAcα-3GlcNAcß-4GlcNAcß-4GlcNAc-DH SU-6 SU-6 SU-6	2,838	3,816	8,524	12,385	8,641
SLeˣ	NeuAcα-3Galß-4GlcNAcß-3Galß-4Glc-DH Fucα-3	3,376	3,120	9,436	17,041	11,970
Sialyl paragloboside	NeuAcα-3Galß-4GlcNAcß-3Galß-4Glcß-Cer	-	-	3,194	3,376	1,222
SLeˣ (Cer)	NeuAcα-3Galß-4GlcNAcß-3Galß-4Glcß-Cer36 Fucα-3	785	542	3,353	2,943	3,123
6SU-SLeˣ (Cer)	SU-6 NeuAcα-3Galß-4GlcNAcß-3Galß-4Glcß-Cer36 Fucα-3	1,906	2,416	6,798	8,575	9,627
VIM-2	NeuAcα-3Galß-4GlcNAcß-3Galß-4GlcNAcß-3Galß-4Glcß-Cer36 Fucα-3	-	-	3,507	6,094	6,669

*-, less than 500.
Expressed on ciliated cells of human bronchial epithelium

Figure 6. Highlights of microarray analyses of pandemic influenza A(H1N1) 2009 (pdm) viruses showing wild-type (222D) pdm viruses binding to α2,3 sialyl sequences with sulphate or fucose at penultimate N-acetylglucosamine, and mutant (222G) viruses increased binding to these and also binding to unmodified or internally fucosylated (VIM-2) sequences. The latter are known to be expressed on ciliated epithelial cells of the human bronchus (adapted from Refs. 56 and 57).

of binding to these was enhanced with mutant viruses (D222G mutation of hemagglutinin) isolated from cases of severe or fatal pdm virus infection: with the D222G mutant pdm viruses there was binding also to nonsulfated internally fucosylated (VIM-2 antigen-bearing) sequences of the type expressed on ciliated cells of the bronchus (Fig. 6).[57] The D222G mutant pdm viruses showed a change in cell tropism in cultures of differentiated human airway epithelial cells; they infected a higher proportion of ciliated epithelial cells than the wild-type pdm viruses. The additional binding of pdm viruses to α2–3 sialyl sequences that occur throughout the airway, including the lung, may account, at least in part, for the capacity of the pdm viruses to cause more severe disease than observed with seasonal influenza. These findings uncover potential mechanisms linking mutation D222G to severity of disease. If the mutant virus were to acquire the ability to spread more widely, the potential consequences would be of great significance, hence the need to closely monitor the evolution of this virus.

The various patterns of recognition of sulfo and fuco motifs by lectins of the immune system and the pathobiolgical consequences of the usurping of these by pathogens such as influenza viruses require investigation.

An increasing variety of carbohydrate microarray systems

At the request of reviewers of the manuscript, I refer briefly to the considerable number of carbohydrate microarray systems that have been developed since 2002 using monosaccharides, polysaccharides, oligosaccharides, and glycolipids as starting materials. These and the diverse immobilization strategies used for the saccharides have been reviewed in detail.[45,48,68–76] Among these are microarray systems with focused repertoires powerfully addressing specific recognition systems. In addition to our NGL-based system (http://www3.imperial.ac.uk/glycosciences), microarray systems with relatively large oligosaccharide coverage to screen for the specificities of diverse carbohydrate-binding systems are those of the Consortium of Functional Glycomics (CFG), Paulson, Cummings Bovin, and Gildersleeve reviewed in Ref 48; see also http://www.functionalglycomics.org/static/consortium/consortium.shtml. There is also the shotgun glycomics approach of Cummings and Smith for generating and probing microarrays from glycomes.[77] Microarray analyses are variously generating an enormous amount of infomation on diverse carbohydrate-recognition systems. Without a doubt, glycobiologists will increasingly establish own microarray systems.

Clearly, the time is ripe to set up studies to formally compare the major microarray platforms that are being used to provide screening analysis data to the wide biomedical community. A cursory overview of results from the NGL-based microarray system and those from the CFG indicates that with robust carbohydrate-binding systems, there is good agreement. But there are differences that we are aware of, for example, the ability of the NGL system uniquely to detect binding of the 2009 pdm influenza virus (wild type), to α2,3 linked sialyl sequences.[56] Are the differences a reflection of the analysis platforms or of the virus preparations used in the analyses? Platform comparisons need to be well orchestrated. The selection of recognition systems for the comparative studies needs to be based on a critical overview of existing data. This would form the basis for having preparations of particular proteins to share for the parallel analyses. There is also a need for standardization of experiments. This is now one of the missions of the project MIRAGE, which stands for minimum information required for a glycomics experiment (http://glycomics.ccrc.uga.edu/MIRAGE/index.php/Main_Page). The Interaction Analysis working group therein aims to define checklists for the standardization of experimental glycomics data and *meta* information.

Concluding remarks

A decade after they were introduced, carbohydrate microarrays have come of age as essential tools in the exploration of glycomes for ligands of proteins involved in biological recognition systems. The data (i.e., the hits elicited) are informative and sometimes utterly unexpected. Microarrays are, of course, screening tools; they provide leads to be pursued to determine the biological significance of glycan-binding signals revealed. Without a doubt, microarrays are ideal tools for screening proteins in proteomes for carbohydrate-binding activities and assigning the oligosaccharide ligands within glycomes.

Acknowledgments

The Carbohydrate Microarray Facility is a Wellcome Trust–supported Biomedical Resource for the broad biomedical community. I thank my colleagues in the Glycosciences Laboratory, and many other collaborators, for their contributions to the establishment and application of the neoglycolipid-based microarray system. I also acknowledge other research support by the UK Medical Research Council, the UK Research Councils' joint Basic Technology Initiative Glycoarrays, the Engineering and Physical Sciences Research Council's "Basic Technology Translational Consortium grant Exploitation of Glycoarrays—Translation to End-Use, and the NCI Alliance of Glycobiologists."

Conflicts of interest

The author declares no conflicts of interest.

Supporting Information

Additional supporting information may be found in the online version of this article.

Figure S1. Preparation of probes as neoglycolipids (NGL) from reducing oligosaccharides by reductive amination or oxime ligation, for noncovalent immobilization on matrices: nitrocellulose, silica, plastics.

Figure S2. The neoglycolipid (NGL) technology.

Figure S3. NGL technology: contributions (1985–2002) prior to miniaturizing for microarrays.

Figure S4. Schematic representation of complement glycoprotein C3 and derived glycoproteins and peptides. Approximate molecular masses of the α and β chains, and the fragments derived from the α chain following proteolysis in the complement cascade are indicated. Known glycosylation sites (filled lollipop), Asn-917 of the α chain and Asn-63 of the β chain, and a third nonglycosylated consensus site (unfilled lollipop), Asn-1595 of the α chain, are indicated (taken from Solis *et al.* 1994. *J. Biol. Chem.* **269:** 11555–11562).

Figure S5. Composition of NGL-based microarrays.

Figure S6. NGL-based microarray system: contributions since 2002.

Figure S7. Dectin-1 and antibody 2G8-IgG binding to linear β1–3-linked glucose sequences derived from curdlan polysaccharide.

References

1. Feizi, T. 1985. Demonstration by monoclonal antibodies that carbohydrate structures of glycoproteins and glycolipids are onco-developmental antigens. *Nature* **314:** 53–57.

2. Gagneux, S., K. Deriemer, T. Van, et al. 2006. Variable host-pathogen compatibility in Mycobacterium tuberculosis. Proc. Natl. Acad. Sci. U. S. A **103**: 2869–2873.

3. Varki, A. & P. Gagneux. 2012. Multifarious roles of sialic acids in immunity. Ann. N. Y. Acad. Sci. **1253**: 16–36.

4. Varki, A., R.D. Cummings, J.D. Esko, et al., Eds. 2009. Essentials of Glycobiology, 2nd ed. Cold Spring Harbor Laboratory Press. Cold Spring Harbor, NY; Available at: http://www.ncbi.nlm.nih.gov/books/NBK1908/.

5. Schallus, T., C. Jaeckh, K. Feher, et al. 2008. Malectin— a novel carbohydrate-binding protein of the endoplasmic reticulum and a candidate player in the early steps of protein n-glycosylation. Mol. Biol. Cell **19**: 3404–3414.

6. Campanero-Rhodes, M.A., A. Smith, W. Chai, et al. 2007. N-glycolyl GM1 ganglioside as a receptor for simian virus 40. J. Virol. **81**: 12846–12858.

7. Tang, P.W., H.C. Gooi, M. Hardy, et al. 1985. Novel approach to the study of the antigenicities and receptor functions of carbohydrate chains of glycoproteins. Biochem. Biophys. Res. Commun. **132**: 474–480.

8. Feizi, T., M.S. Stoll, C.-T. Yuen, et al. 1994. Neoglycolipids: probes of oligosaccharide structure, antigenicity and function. Methods Enzymol. **230**: 484–519.

9. Kabat, E.A. 1982. Philip Levine Award Lecture. Contributions of quantitative immunochemistry to knowledge of blood group A, B, H, Le, I and i antigens. Am. J. Clin. Pathol. **78**: 281–292.

10. Watkins, W.M. 1980. Biochemistry and genetics of the ABO, Lewis and P blood group systems. Adv. Hum. Genet. **10**: 1–136, 379–385.

11. Gooi, H.C., T. Feizi, A. Kapadia, et al. 1981. Stage specific embryonic antigen SSEA-1 involves α1–3 fucosylated type 2 blood group chains. Nature **292**: 156–158.

12. Gooi, H.C., S.J. Thorpe, E.F. Hounsell, et al. 1983. Marker of peripheral blood granulocytes and monocytes of man recognized by two monoclonal antibodies VEP8 and VEP9 involves the trisaccharide 3-fucosyl-N-acetyllactosamine. Eur. J. Immunol. **13**: 306–312.

13. Stoll, M.S., T. Mizuochi, R.A. Childs & T. Feizi. 1988. Improved procedure for the construction of neoglycolipids having antigenic and lectin-binding activities from reducing oligosaccharides. Biochem. J. **256**: 661–664.

14. Magnani, J.L., D.F. Smith & V. Ginsburg. 1980. Detection of gangliosides that bind cholera toxin: direct binding of I-labeled toxin to thin-layer chromatograms. Anal. Biochem. **109**: 399–402.

15. Chai, W., M.S. Stoll, C. Galustian, et al. 2003. Neoglycolipid technology—deciphering information content of glycome. Methods Enzymol. **362**: 160–195.

16. Liu, Y., T. Feizi, M.A. Campanero-Rhodes, et al. 2007. Neoglycolipid probes prepared via oxime ligation for microarray analysis of oligosaccharide-protein interactions. Chem. Biol. **14**: 847–859.

17. Yuen, C.-T., A.M. Lawson, W. Chai, et al. 1992. Novel sulfated ligands for the cell adhesion molecule E-selectin revealed by the neoglycolipid technology among O-linked oligosaccharides on an ovarian cystadenoma glycoprotein. Biochemistry **31**: 9126–9131.

18. Streit, A., C.-T. Yuen, R.W. Loveless, et al. 1996. The Le^x carbohydrate sequence is recognized by antibody to L5, a functional antigen in early neural development. J. Neurochem. **66**: 834–844.

19. Leteux, C., W. Chai, K. Nagai, et al. 2001. 10E4 Antigen of scrapie lesions contains an unusual nonsulfated heparan motif. J. Biol. Chem. **276**: 12539–12545.

20. Rosenstein, I.J., M.S. Stoll, T. Mizuochi, et al. 1988. Oligosaccharide probes reveal a new type of adhesive specificity in Escherichia coli from patients with urinary tract infection. Lancet **2**: 1327–1330.

21. Yuen, C.-T., W. Chai, R.W. Loveless, et al. 1997. Brain contains HNK-1 immunoreactive O-glycans of the sulfoglucuronyl lactosamine series that terminate in 2-linked or 2,6-linked hexose (mannose). J. Biol. Chem. **272**: 8924–8931.

22. Chai, W., C.T. Yuen, H. Kogelberg, et al. 1999. High prevalence of 2-mono- and 2,6-di-substituted Manol-terminating sequences among O-glycans released from brain glycopeptides by reductive alkaline hydrolysis. Eur. J. Biochem. **263**: 879–888.

23. Leteux, C., W. Chai, R.W. Loveless, et al. 2000. The cysteine-rich domain of the macrophage mannose receptor is a multispecific lectin that recognizes chondroitin sulfates A and B and sulfated oligosaccharides of blood group Lewis^a and Lewis^x types in addition to the sulfated N-glycans of lutropin. J. Exp. Med. **191**: 1117–1126.

24. Loveless, R.W., T. Feizi, R.A. Childs, et al. 1989. Bovine serum conglutinin is a lectin which binds non-reducing terminal N-acetylglucosamine, mannose and fucose residues. Biochem. J. **258**: 109–113.

25. Mizuochi, T., R.W. Loveless, A.M. Lawson, et al. 1989. A library of oligosaccharide probes (neoglycolipids) from N-glycosylated proteins reveals that conglutinin binds to certain complex type as well as high-mannose type oligosaccharide chains. J. Biol. Chem. **264**: 13834–13839.

26. Solis, D., T. Feizi, C.T. Yuen, et al. 1994. Differential recognition by conglutinin and mannan-binding protein of N-glycans presented on neoglycolipids and glycoproteins with special reference to complement glycoprotein C3 and ribonuclease. J. Biol. Chem. **269**: 11555–11562.

27. Childs, R.A., K. Drickamer, T. Kawasaki, et al. 1989. Neoglycolipids as probes of oligosaccharide recognition by recombinant and natural mannose-binding proteins of the rat and man. Biochem. J. **262**: 131–138.

28. Childs, R.A., T. Feizi, K. Drickamer & M.S. Quesenberry. 1990. Differential recognition of core and terminal portions of oligosaccharide ligands by carbohydrate-recognition domains of two mannose-binding proteins. J. Biol. Chem. **265**: 20770–20777.

29. Loveless, R.W., U. Holmskov & T. Feizi. 1995. Collectin-43 is a serum lectin with a distinct pattern of carbohydrate recognition. Immunology **85**: 651–659.

30. Childs, R.A., J.R. Wright, G.F. Ross, et al. 1992. Specificity of lung surfactant protein SP-A for both the carbohydrate and the lipid moieties of certain neutral glycolipids. J. Biol. Chem. **267**: 9972–9979.

31. Loveless, R.W., G. Floyd-O'sullivan, J.G. Raynes & T. Feizi. 1992. Human serum amyloid P is a multispecific adhesive protein whose ligands include 6-phosphorylated mannose and the 3-sulphated saccharides galactose, N-acetylgalactosamine and glucuronic acid. EMBO J. **11**: 813–819.

32. Larkin, M., T.J. Ahern, M.S. Stoll, *et al.* 1992. Spectrum of sialylated and non-sialylated fuco-oligosaccharides bound by the endothelial-leukocyte adhesion molecule E-selectin. Dependence of the carbohydrate binding activity on E-selectin density. *J. Biol. Chem.* **267:** 13661–13668.

33. Green, P.J., T. Tamatani, T. Watanabe, *et al.* 1992. High affinity binding of the leucocyte adhesion molecule L-selectin to 3'-sulphated-Le^a and -Le^x oligosaccharides and the predominance of sulphate in this interaction demonstrated by binding studies with a series of lipid-linked oligosaccharides. *Biochem. Biophys. Res. Commun.* **188:** 244–251.

34. Green, P.J., C.-T. Yuen, R.A. Childs, *et al.* 1995. Further studies of the binding specificity of the leukocyte adhesion molecule, L-selectin, towards sulphated oligosaccharides—suggestion of a link between the selectin- and the integrin-mediated lymphocyte adhesion systems. *Glycobiology* **5:** 29–38.

35. Osanai, T., T. Feizi, W. Chai, *et al.* 1996. Two families of murine carbohydrate ligands for E-selectin. *Biochem. Biophys. Res. Commun.* **218:** 610–615.

36. Galustian, C., A. Lubineau, C. Le Narvor, *et al.* 1999. L-selectin interactions with novel mono- and multisulfated Lewis^x sequences in comparison with the potent ligand 3'-sulfated Lewis^a. *J. Biol. Chem.* **274:** 18213–18217.

37. Alon, R., T. Feizi, C.-T. Yuen, *et al.* 1995. Glycolipid ligands for selectins support leukocyte tethering and rolling under physiologic flow conditions. *J. Immunol.* **154:** 5356–5366.

38. Solomon, J.C., M.S. Stoll, P. Penfold, *et al.* 1991. Studies of the binding specificity of the soluble 14 000-dalton bovine heart muscle lectin using immobilised glycolipids and neoglycolipids. *Carbohydr. Res.* **213:** 293–307.

39. Feizi, T., J.C. Solomon, K.C.G. Jeng, *et al.* 1994. The adhesive specificity of the soluble human lectin, IgE-binding protein, toward lipid-linked oligosaccharides. Presence of blood group A, B, B-like, and H monosaccharides confers a binding activity to tetrasaccharide (Lacto-*N*-tetraose and Lacto-*N*-neotetraose) backbones. *Biochemistry* **33:** 6342–6349.

40. Galustian, C., R.A. Childs, M.S. Stoll, *et al.* 2002. Synergistic interactions of the two classes of ligand, sialyl-Lewis$^{a/x}$ fuco-oligosaccharides and short sulpho-motifs, with the P- and L- selectins: implications for therapeutic inhibitor designs. *Immunology* **105:** 350–359.

41. Lawson, A.M., W. Chai, G.C. Cashmore, *et al.* 1990. High-sensitivity structural analyses of oligosaccharide probes (neoglycolipids) by liquid-secondary-ion mass spectrometry. *Carbohydr. Res.* **200:** 47–57.

42. Palma, A.S., T. Feizi, Y. Zhang, *et al.* 2006. Ligands for the beta-glucan receptor, Dectin-1, assigned using 'designer' microarrays of oligosaccharide probes (neoglycolipids) generated from glucan polysaccharides. *J Biol. Chem.* **281:** 5771–5779.

43. Feizi, T. & W. Chai. 2004. Oligosaccharide microarrays to decipher the glyco code. *Nat. Rev. Mol. Cell Biol.* **5:** 582–588.

44. Fukui, S., T. Feizi, C. Galustian, *et al.* 2002. Oligosaccharide microarrays for high-throughput detection and specificity assignments of carbohydrate-protein interactions. *Nat. Biotechnol.* **20:** 1011–1017.

45. Liu, Y., A.S. Palma & T. Feizi. 2009. Carbohydrate microarrays: key developments in glycobiology. *Biol. Chem.* **390:** 647–656.

46. Palma, A.S., Y. Liu, R.A. Childs, *et al.* 2011. The human epithelial carcinoma antigen recognized by monoclonal antibody AE3 is expressed on a sulfoglycolipid in addition to neoplastic mucins. *Biochem. Biophys. Res. Commun.* **408:** 548–552.

47. Stoll, M.S. & T. Feizi. 2009. Software tools for storing, processing and displaying carbohydrate microarray data. In *Proceeding of the Beilstein Symposium on Glyco-Bioinformatics*, 4–8 October, 2009. Potsdam, Germany. C. Kettner, Ed.: 123–140.

48. Rillahan, C.D. & J.C. Paulson. 2011. Glycan microarrays for decoding the glycome. *Annu. Rev. Biochem.* **80:** 797–823.

49. Galustian, C., C.G. Park, W. Chai, *et al.* 2004. High and low affinity carbohydrate ligands revealed for murine SIGN-R1 by carbohydrate array and cell binding approaches, and differing specificities for SIGN-R3 and langerin. *Int. Immunol.* **16:** 853–866.

50. Reddy, S.T., W. Chai, R.A. Childs, *et al.* 2004. Identification of a low affinity mannose 6-phosphate-binding site in domain 5 of the cation-independent mannose 6-phosphate receptor. *J. Biol. Chem.* **279:** 38658–38667.

51. Campanero-Rhodes, M.A., R.A. Childs, M. Kiso, *et al.* 2006. Carbohydrate microarrays reveal sulphation as a modulator of siglec binding. *Biochem. Biophys. Res. Comm.* **344:** 1141–1146.

52. Liu, Y., A.S. Palma, Y. Zhang, *et al.* 2007. Oxime-linked neoglycolipids—powerful tools for recognition studies of glucan-binding proteins in microarrays [Conference Abstract]. *Glycobiology* **17:** 1257.

53. Redelinghuys, P., A. Antonopoulos, Y. Liu, *et al.* 2011. Early murine T-lymphocyte activation is accompanied by a switch from N-glycolyl- to N-acetyl-neuraminic acid and generation of ligands for siglec-E. *J. Biol. Chem.* **286:** 34522–34532.

54. Blumenschein, T.M.A., N. Friedrich, R.A. Childs, *et al.* 2007. Atomic resolution insight into host cell recognition by *Toxoplasma gondii*. *EMBO J.* **26:** 2808–2820.

55. Garnett, J.A., Y. Liu, E. Leon, *et al.* 2009. Detailed insights from microarray and crystallographic studies into carbohydrate recognition by microneme protein 1 (MIC1) of Toxoplasma gondii. *Protein Sci.* **18:** 1935–1947.

56. Childs, R.A., A.S. Palma, S. Wharton, *et al.* 2009. Receptor-binding specificity of pandemic influenza A (H1N1) 2009 virus determined by carbohydrate microarray. *Nat. Biotechnol.* **27:** 797–799.

57. Liu, Y., R.A. Childs, T. Matrosovich, *et al.* 2010. Altered receptor specificity and cell tropism of D222G hemagglutinin mutants isolated from fatal cases of pandemic A(H1N1) 2009 influenza virus. *J. Virol.* **84:** 12069–12074.

58. Torosantucci, A., P. Chiani, C. Bromuro, *et al.* 2009. Protection by anti-beta-glucan antibodies is associated with restricted beta-1,3 glucan binding specificity and inhibition of fungal growth and adherence. *PLoS. ONE* **4:** e5392.

59. Capodicasa, C., P. Chiani, C. Bromuro, *et al.* 2011. Plant production of anti-beta-glucan antibodies for immunotherapy of fungal infections in humans. *Plant Biotechnol. J.* **9:** 776–787.

60. Pejchal, R., K.J. Doores, L.M. Walker, *et al.* 2011. A potent and broad neutralizing antibody recognizes and penetrates the HIV glycan shield. *Science* **334:** 1097–1103.

61. Mouquet, H., L. Scharf, Z. Euler, *et al.* 2012. Complex-type *N*-glycan recognition by potent broadly neutralizing HIV antibodies. *Proc. Natl. Acad. Sci. U. S. A* **109:** E3268–E3277.

62. Brown, G.D., P.R. Taylor, D.M. Reid, *et al.* 2002. Dectin-1 is a major beta-glucan receptor on macrophages. *J. Exp. Med.* **196:** 407–412.

63. Brown, G.D. & S. Gordon. 2001. Immune recognition. A new receptor for beta-glucans. *Nature* **413:** 36–37.

64. Brisson, J.R., H. Baumann, A. Imberty, *et al.* 1992. Helical epitope of the group B meningococcal alpha(2–8)-linked sialic acid polysaccharide. *Biochemistry* **31:** 4996–5004.

65. Kogelberg, H. & T.J. Rutherford. 1994. Studies on the three-dimensional behaviour of the selectin ligands Lewis(a) and sulphated Lewis(a) using NMR spectroscopy and molecular dynamics simulations. *Glycobiology* **4:** 49–57.

66. Galustian, C., A.M. Lawson, S. Komba, *et al.* 1997. Sialyl-Lewis[x] sequence 6-*O*-sulfated at *N*-acetylglucosamine rather than at galactose is the preferred ligand for L-selectin and de-*N*-acetylation of the sialic acid enhances the binding strength. *Biochem. Biophys. Res. Comm.* **240:** 748–751.

67. Kawashima, H. & M. Fukuda. 2012. Sulfated glycans control lymphocyte homing. *Ann. N. Y. Acad. Sci.* **1253:** 112–121.

68. Feizi, T., F. Fazio, W. Chai & C.H. Wong. 2003. Carbohydrate microarrays—a new set of technologies at the frontiers of glycomics. *Curr. Opin. Struct. Biol.* **13:** 637–645.

69. Liang, P.H., C.Y. Wu, W.A. Greenberg & C.H. Wong. 2008. Glycan arrays: biological and medical applications. *Curr. Opin. Chem. Biol.* **12:** 86–92.

70. Park, S., M.R. Lee & I. Shin. 2009. Construction of carbohydrate microarrays by using one-step, direct immobilizations of diverse unmodified glycans on solid surfaces. *Bioconj. Chem.* **20:** 155–162.

71. Fangel, J.U., H.L. Pedersen, S. Vidal-Melgosa, *et al.* 2012. Carbohydrate microarrays in plant science. *Methods Mol. Biol.* **918:** 351–362.

72. Oyelaran, O. & J.C. Gildersleeve. 2009. Glycan arrays: recent advances and future challenges. *Curr. Opin. Chem. Biol.* **13:** 406–413.

73. Lepenies, B. & P.H. Seeberger. 2010. The promise of glycomics, glycan arrays and carbohydrate-based vaccines. *Immunopharmacol. Immunotoxicol.* **32:** 196–207.

74. Liang, C.H. & C.Y. Wu. 2009. Glycan array: a powerful tool for glycomics studies. *Expert. Rev. Proteomics* **6:** 631–645.

75. Smith, D.F., X. Song & R.D. Cummings. 2010. Use of glycan microarrays to explore specificity of glycan-binding proteins. *Methods Enzymol.* **480:** 417–444.

76. Gao, J., L. Ma, D. Liu & Z. Wang. 2012. Microarray-based technology for glycomics analysis. *Comb. Chem. High Throughput Screen* **15:** 90–99.

77. Song, X., Y. Lasanajak, B. Xia, *et al.* 2011. Shotgun glycomics: a microarray strategy for functional glycomics. *Nat. Methods* **8:** 85–90.